Praise for *Street Far*

"*Street Farm* is a story of how to bring cities back to life, literally and emotionally. The cold, forbidding landscapes of urban life bring our hearts to a standstill. When streets, medians, abandoned land, parks, and byways are transformed by soil, bugs, microbes, pollinators, and seeds, lives bloom. Connectedness flourishes, and people become denizens once again.

"Local food is not a mere talisman or gesture. Local food not only addresses quality of life, economy, and food security, it changes our hearts. Michael Ableman has a finely honed sensibility. Read how he gardens society, grows well-being, weeds out despair, and sows hope in this wonderfully written testament to life."

—**PAUL HAWKEN**, author of *Blessed Unrest*

"Whenever Michael Ableman sees a barrier, he runs over and kicks it in. Lucky for us, this strikingly focused anarchist writes about it too, sharing the deeply moving story of reclaiming land and building real community in the most unlikely places, from the ground up. Read this book and be amazed."

—**DAN BARBER**, chef/co-owner, Blue Hill; author of *The Third Plate*

"Michael Ableman is one of the pioneers of small-scale urban farming, growing quality food for urban communities. He has worked through the challenges inherent to urban farming. Michael has been and is an inspiration to myself and many urban agriculture leaders around the country and the world."

—**WILL ALLEN**, founder and CEO, Growing Power

"This is the most inspiring book I have read in years. I found myself trembling at the monumental challenges that Michael Ableman and his colleagues faced and overcame in creating a set of urban farms in some of the most downtrodden neighborhoods on the continent. This is a story of hope, disappointment, and hope returning, detailing the mistakes and setbacks as well as the victories and benefits of creating a large-scale food-growing program in a big city. Told in moving vignettes and full of useful tips for those who want to try to heal the urban food grid, this is an important book. It's essential reading for everyone in the urban food movement."

—**TOBY HEMENWAY**, author of *The Permaculture City* and *Gaia's Garden*

"In *Street Farm*, long-time farmer Michael Ableman reports on the triumphs and failures of Vancouver's Sole Food Street Farms. The goal of this five-acre network of four farms—begun in the poorest postal code in Canada—is to produce, from thousands of boxes of planted dirt, not just delicious food but salvaged lives. Candid about the difficulties of creating flourishing farms on hot pavements and of making reliable farm workers of dispirited locals who struggle not only with poverty but with assorted personal demons, Ableman has written an important, inspiring, and bravely honest book."

—**JOAN GUSSOW**, author of *Growing, Older* and *This Organic Life*

"Michael Ableman is an innovator extraordinaire whose projects have a track record of benchmarking new models of best practice. He is one of the handful of inspiring visionaries on the planet who are redefining our future food systems."

—PATRICK HOLDEN, founding director, Sustainable Food Trust

"In this inspiring book, Michael Ableman documents that generating paradise by growing vegetables amidst the urban jungle also rehabilitates lost souls, builds community, and creates genuine economic value. *Street Farm* is a great antidote to pessimism, illustrating how even seemingly broken people can contribute to themselves, to society, and to our shared ecology."

—GABOR MATÉ, MD, author of *In the Realm of Hungry Ghosts*

"Michael Ableman examines the heart and soul of urban agriculture through the eyes, hands, and hearts of people in need of a place of civility and serenity. The passion and humility of the farmers who work at Sole Food Street Farms shines through. They are neighborhood folks, many with transgressions of addictions, who find solace in farming. From *Street Farm*, we learn that urban agriculture indeed takes a village of planners, politicians, investors, and believers to envision such an economy, with urban agriculture as the new economic engine."

—KAREN WASHINGTON, urban farm activist;
cofounder of Black Urban Growers

"Sole Food Street Farms is living proof that creative social enterprises, thoughtful land use, and green jobs can combine to make cities more inclusive and resilient. Michael Ableman's work and passion helped make Vancouver a global leader in urban food systems, with happier and healthier people."

—GREGOR ROBERTSON, mayor, Vancouver, British Columbia

"Michael Ableman recognises that urban growing is not just about producing lovely, healthy, local food. It's about creating meaningful work that pays a decent living and showing that cities can play a vital role in building a better, more resilient food system. In *Street Farm*, Ableman writes about many of the issues that we also grapple with as we strive to build a better food system in London. Sole Food Street Farms is an uplifting demonstration of how communities really can change the world."

—JULIE BROWN, director, Growing Communities

"Michael Ableman's interwoven growing skills and people empowerment are beautifully illustrated by 'ground zero' spaces transformed to market gardens. Sole Food Street Farms produces twenty-five tons of food every year, grown in unlikely places by drug-addicted farmers, softened in the process like the soil they tend."

—CHARLES DOWDING, author of *How to Create a New Vegetable Garden*

"*Street Farm* tells it like it is on a gritty urban farm, introducing us to rough but real people who learn to live again through growing food and nurturing the soil. Michael Ableman shows us that we can amend distressed soils and distressed communities alike." —NOVELLA CARPENTER, author of *Farm City*

STREET FARM

STREET FARM

Growing Food, Jobs, and Hope on the Urban Frontier

MICHAEL ABLEMAN

Chelsea Green Publishing
White River Junction, Vermont

Project Manager: Patricia Stone
Acquisitions Editor: Ben Watson
Developmental Editor: Fern Marshall Bradley
Copy Editor: Laura Jorstad
Proofreader: Eileen M. Clawson
Designer: Melissa Jacobson

Printed in the United States of America.
First printing July, 2016.
10 9 8 7 6 5 4 3 2 1   16 17 18 19 20

Our Commitment to Green Publishing

Chelsea Green sees publishing as a tool for cultural change and ecological stewardship. We strive to align our book manufacturing practices with our editorial mission and to reduce the impact of our business enterprise in the environment. We print our books and catalogs on chlorine-free recycled paper, using vegetable-based inks whenever possible. This book may cost slightly more because it was printed on paper that contains recycled fiber, and we hope you'll agree that it's worth it. Chelsea Green is a member of the Green Press Initiative (www.greenpressinitiative.org), a nonprofit coalition of publishers, manufacturers, and authors working to protect the world's endangered forests and conserve natural resources. *Street Farm* was printed on paper supplied by RR Donnelley that contains at least 10 percent postconsumer recycled fiber.

Library of Congress Cataloging-in-Publication Data
Names: Ableman, Michael, author.
Title: Street farm : growing food, jobs, and hope on the urban frontier / Michael Ableman.
Description: White River Junction, Vermont : Chelsea Green Publishing, [2016]
Identifiers: LCCN 2016017618 | ISBN 9781603586023 (pbk.) | ISBN 9781603586030 (ebook)
Subjects: LCSH: Urban agriculture—British Columbia—Vancouver—Case studies.
Classification: LCC S451.5.B65 A25 2016 | DDC 338.109711/28—dc23
LC record available at https://lccn.loc.gov/2016017618

Chelsea Green Publishing
85 North Main Street, Suite 120
White River Junction, VT 05001
(802) 295-6300
www.chelseagreen.com

To the memory of my father, to my mother,
to my wife, Jeanne-Marie, and
my two sons, Benjamin and Aaron, and
to every member of Sole Food's Downtown Eastside crew,
who have inspired me with their courage and perseverance.

Also by Michael Ableman

Fields of Plenty:
A Farmer's Journey in Search of
Real Food and the People Who Grow It

On Good Land:
The Autobiography of an Urban Farm

From the Good Earth:
A Celebration of Growing
Food Around the World

Contents

———

Sometimes it is the people no one imagines anything of who do the things that no one can imagine.

—*The Imitation Game*

Introduction

In the fall of 2009 I received a phone call asking me if I would come to Vancouver to attend a meeting. At the time I was living on Salt Spring Island, off the coast of British Columbia, running my family farm, known as Foxglove Farm. "It would only require a couple hours of your time," I was told. "We want to share an idea, draw on your experience." I hesitated, feeling that familiar stirring when a new project is about to begin, knowing all too well how one two-hour meeting can lead to a lifetime of work.

When I received this call, I had been farming full-time for almost forty years. In the mid-1980s I founded the Center for Urban Agriculture, and through that nonprofit I'd been involved with farms in some of the most challenged neighborhoods in North America. I agreed to attend that meeting with folks from several nonprofits and social service organizations working on the city's Downtown Eastside. They wanted to talk about growing fresh food and jobs for a neighborhood that needed both.

One meeting led to another, and before long I'd partnered with Seann Dory, who was working with United We Can, a local organization that employed people from the neighborhood to clean up the streets and alleys and to recycle cans and bottles. We began with a half-acre parking lot, but even then I imagined farms throughout Vancouver, places where disenfranchised people could learn new skills, participate in meaningful work, maybe heal themselves, and help feed the city.

Street Farm is the story of Sole Food Street Farms, a network of four urban farms located on five acres of reclaimed land throughout downtown Vancouver. We produce over twenty-five tons of fresh produce per year, including tree fruit from a large urban orchard, supply more than thirty area restaurants, sell at five Vancouver farmers markets, and operate a community supported agriculture program. We also donate up to $20,000 of produce per year to community kitchens and provide jobs to twenty-five people. Central to our vision from the beginning has been a commitment to building a community with and for the people we've hired and trained—among them the poor and homeless, the drug-addicted and mentally ill—and the story of the farm is as much about the farmers I've come to work with as it is the food we've produced together.

The narrative you're about to read recounts the founding and growth—in all its fits and starts—of this remarkable agricultural project. We've done good work, I think. We've made mistakes. You'll see all that in what's to come.

Throughout my working life, I've found it's impossible to maintain the energy and the will a project like Sole Food requires without the belief that it will create positive change. But big ideas about what other people need can be dangerous. And I've learned from the people I work with to adjust my expectations and keep in check my illusions (even those I hold about myself), remembering that our little effort cannot resolve the core challenges that exist for those of inner-city Vancouver.

In each chapter that follows you'll meet some of the farmers I've worked with, some of the lives we've tried to reach by filling planters with soil and growing good food. Photographs of many of them appear throughout the book. This is their story as much as it is mine, and the success of our farms, which has been a profound inspiration to me, I owe to them.

Throughout *Street Farm* you'll also find mini essays that bring together some of what I've learned over the last four decades about farming and urban agriculture: the soil and water, weeding and cultivation, our use of time and space, the importance of walking the land, and more. All of these lessons have helped shape Sole Food in one way or another, and as you'll see I often find myself returning to them as I welcome new farmers into our community or remind those who have been with us for some time about the range of skills, the knowledge, the art, and the compassion it takes to feed a neighborhood.

CHAPTER ONE

Astoria

"Food's the next thing, man!"

I turn to see a local Hells Angel yelling to no one in particular, shaking the chain-link fence at our corner of Hawks and Hastings in Vancouver. He's looking into the parking lot where I'm working, now a verdant urban farm.

The street where he's standing in his tattoos and leather runs elevated above our half acre. Passersby cannot help but look down into the incongruent goings-on at Sole Food Street Farms. Every day a few among the throngs stop to gawk in amazement at what had long been an unsightly expanse of asphalt but is now a full-on agricultural enterprise bursting with crops, smack in the middle of one of the most infamous neighborhoods in the world, Vancouver's Downtown Eastside, the Low Track, ground zero, home of the term *skid row*. Ours is the poorest postal code in the country and holds the dubious distinction of having the highest rates of HIV and hepatitis C per capita in North America. It is also home to one of the continent's highest concentrations of open prostitution.

And if food is the next thing, man, drugs have been the thing until now. In fact, to say they've moved on would be wishful thinking. Wind your way through the neighborhood between Carrall and Columbia Streets on East Hastings, nearby, and you hear "rock, powder, down" repeated over and over, to no one in particular. It's a mantra in an open offering of crack, powder cocaine, and heroin to everyone who walks by. And it kills; there are more drug-related deaths here than anywhere else on the continent. Every kind of dope is available here, along with an amazing array of stolen bicycles, suitcases, cell phones, clothing, shoes. And in a back alley, a cheap hotel room, or the comfort of your car, any number of illicit sexual experiences are on offer.

All of this squalor exists in the heart of gleaming steel-and-glass high-rises, pristine parks, art galleries, and restaurants, as the shiny new Mercedes-Benzes, Lexuses, and BMWs ply the streets of what is now considered North America's most expensive city. While Vancouver's prosperity is celebrated, its concentration of poverty and raw desperation endures in the midst of the polished and the preened.

The lot where our farm sits is attached to the Astoria Hotel. Built in 1912, the Astoria was once a prominent feature in this neighborhood, but like the neighborhood that surrounds it, the building has drifted into aging and decay. Now the Astoria is one of several hotels owned by one of the city's most notorious slumlords. For $425 per month you can get an eight-by-ten single room on one of five floors of the hotel. The rooms are rough: stained carpets, peeling paint, droves of bedbugs, and an ongoing cacophony of raw, uncensored life filtering through the walls. Pig farmer and convicted mass murderer Robert Pickton hung out at this hotel, reportedly abducting prostitutes whom he later murdered at his farm outside the city. Thirty-two women from this neighborhood were killed by Pickton. Some were last seen at the Astoria.

Still today, girls as young as thirteen wearing black fishnet stockings, thigh-high tight red skirts, high heels, and makeup solicit johns on the corner by Sole Food Farm only feet away from where we seed our spinach or trellis tomatoes or harvest beans. Walk or drive the alleys behind the buildings on the main drag and people are injecting and inhaling, urinating in the street, and, at night, openly fucking.

Come around on Welfare Wednesday, known as Mardi Gras to the locals, and a line of people extends out onto the street from the Astoria's discount beer and wine store. The hotel is also home to a renowned bar. Once it served loggers, miners, and fishermen whose seasonal work brought them to the boardinghouses and watering holes. In recent years it's been discovered by local hipsters who pride themselves on venturing into this forbidden neighborhood to get cheap drinks and listen to bands like 3 Inches of Blood, Dayglo Abortions, the Mutators, and the Japandroids.

Our farm staff uses the basement bathroom in the Astoria. We climb the stairs from the farm to the street and head through the liquor store, down to the basement, and through pallets stacked with Molson, Bud, Coors, High Test, and Colt, past a full-scale boxing ring, to get to the makeshift door that opens into the single-stall toilet.

On days when it's too early and the liquor store is closed, I hesitantly climb the stairs of the hotel to the bathrooms shared by the single-room boarders. I never know what I'll encounter in the upper reaches of the

hotel—someone sick in the toilet, somebody yelling obscenities, some offer of sex or drugs. Whatever it is, it's always a wake-up call.

Through all this, I'm reminded that although there were other places within and around Vancouver to grow food and that other neighborhoods might have been easier environments for establishing a farm, this one is home for most of our farm crew. And no other site would have been as symbolic as the Astoria.

———————

My only exposure to this neighborhood had been driving down Hastings Street, a major east–west traffic corridor, on my way to and from the Fraser Valley to pick up farm supplies. I never imagined I'd be working here. Sitting in those first meetings, talking with some of the folks who

make this neighborhood their home, hearing about the desperate desire for a new approach to helping people—something other than methadone, clean needles, street nurses, and welfare checks—was sobering. I had no illusions, not one naive thought that I had any answers, just a familiar pull to use my skills to provide good work, create a little beauty, grow something good to eat.

From Sole Food's inception in 2009, my colleague Seann Dory and I saw farming in Vancouver as a powerful way to provide good jobs to people living in this neighborhood. When Seann and I met, he had lived in the city for seven years. I'd come to this part of Canada eight years earlier, establishing my family farm, Foxglove, on Salt Spring Island (situated between the mainland and Vancouver Island) after almost thirty years of farming and activism in California. Island life is wonderful, and I love my friends and neighbors, but the island is a remote and privileged place where the urgent problems in the world can seem far away. When my family and I moved to the island, I thought I wanted to retreat, but within a few years I missed being in the trenches. Vancouver—and now its Downtown Eastside, in particular—has become a place where I feel that I can contribute.

With the help of United We Can, the charitable organization that first supported the work of Sole Food, Seann and I would transform vacant urban land into street farms and grow artisan-quality fruits and vegetables. We would empower individuals with limited resources, many of whom are also managing addiction and chronic mental health problems, by providing jobs, agricultural training, and inclusion in a supportive community of farmers and food lovers.

We wanted our first farm to be in the heart of the neighborhood, located where those we would employ actually lived. Sole Food had to be accessible and visible, because we wanted the broader community to see that there was something other than the hardness of pavement, the stained teeth and hollowed eyes, the hopelessness, the drugs and theft and raw desperation all around. We wanted the world to know that people from this neighborhood, those who were viewed as low-life losers, could create something beautiful and productive; that they could eat from it, feed others, and get a paycheck from its abundance; and that it could sustain itself for more than a few days or weeks or months or years.

The name Sole Food arose from one of United We Can's initiatives: Save Our Living Environment (SOLE). (Although we have since moved out from under the United We Can umbrella, the name remains with us.) Only seven years after starting our first street farm, we are growing on multiple sites in the city, and still maintain our farm in Vancouver's infamous Downtown Eastside, still in the shadow of the Astoria.

The chain-link fence that the Hells Angel shook was not always intact. Not long ago, it was broken and incomplete, an open invitation into the dark back corner of the lot, a place to do drugs or turn tricks. The lot also served as parking for Astoria Hotel's liquor store and a place to illegally dump construction waste and garbage.

Yet in the midst of the trash and rubble and decay, a huge, healthy potato-like plant thrived, bursting out from a crack in the hotel's plaster wall, a hopeful sign of what was to come.

STREET FARM

In the forty years I've been farming, the vast majority of the organically grown food I've produced has been available only to a narrow segment of society: those who can afford it. Even as I've worked to address this basic problem, I've lived for most of my career with some related pressing questions: How can we make high-quality fresh food more affordable and available to all, while still supporting the farmers who work so hard to produce it? Could farming be used to provide jobs and healing to people who have become marginalized because of poverty, mental illness, and drug addiction? Knowing that these social ills are often concentrated in urban centers, I've asked myself: Is it possible to create viable farming enterprises on pavement and contaminated land in the heart of our cities?

During the years I was starting to farm, abandoned land was common in most low-income neighborhoods in North America's cities. Common, too, in these areas were high unemployment and a shortage of fresh food. I founded the Center for Urban Agriculture in the 1980s with the idea that we could create small farm businesses on urban land, providing much-needed employment and food for those underserved communities. Most people at that time were confused when I used the words *urban* and *agriculture* in the same sentence.

Over the years I created and operated both rural and urban farming enterprises. Some looked like the prototypical vision of a farm: wide-open fields, rows of vegetables or fruit trees, a welcoming farmhouse, a large barn, animals grazing, tractors cultivating, and trucks stacked with boxes of food. But some of the projects I started were in urban places like Watts in Los Angeles, a neighborhood known for its poverty, violence, and unemployment. Starting in 1999, through the Center for Urban Agriculture, we created a three-acre farm on the site of the Watts health clinic, which had burned down during the uprisings in the mid-1960s.

I was young when I helped to develop that farm in Watts. Naively, I thought I could cure some of the neighborhood's ills. Like so many do-gooders who had come and gone through there, I believed I had some answers to the deep-seated challenges that plagued that community. I discovered that I knew nothing. My own privilege kept me from really understanding the needs of the people who lived there. I did not come from that neighborhood, had never had to cope with that level of poverty and desperation, had always had a place to live and food to eat, and the color of my skin is not black or brown.

In the end I had to be satisfied that the best I could do was to share some of my skills with the people who call Watts home. Now in my mid-fifties, my idealism tempered but my belief in the power of growing food still intact, I've started another urban farm, in another infamous neighborhood.

Seann and I had been introduced in January 2009 by Liz Charnya, who from her employment services office in Chinatown was organizing training and job opportunities for people from the Downtown Eastside. Seann had already been working as a project manager for United We Can, which employs hundreds of folks from this neighborhood as "binners," cleaning Vancouver's alleys and streets, and running the largest recycling center in the region. Our partnership in this seemingly quixotic endeavor has been built on his youth and energy coupled with my farming experience.

Born in Edmonton, Alberta, in 1977, to working-class parents, Seann embodies a blend of street-smart edge with a deep concern for other people. Seann studied acting, got a few bit parts on *Battlestar Galactica* and in Robert Redford's *Unfinished Life*, and then spent a few years as a bartender and running nightclubs. End up in jail or in the hospital or flat-out broke, and Seann is the one you want to call on. And although we come from very different backgrounds and are more than twenty years apart, together we've been able to raise more than a couple of million dollars, and cobble together a parking lot and a ragtag group of folks from the neighborhood to grow food and jobs and provide some hope and stability.

My work in Watts and my recent urban farming in Vancouver have left me with few illusions about the challenges we face in our effort to help grow these communities. I've shared my experiences with Seann; he's done the same for me. And we sometimes encounter the reality that people are no easier to recover than the land buried under layers of pavement. Ours remains an imperfect endeavor.

But I know that we all need to eat and that we all need something to do each day that connects us to one another and to the broader world around us. If growing fruits and vegetables at Sole Food Street Farms could provide neighborhood men and women a rock-solid, always-there, grounded-in-the-ground place to return to, then we would accomplish a great deal. Producing excellent-quality food for the community and creating a lush, green, multidimensional break from so much pavement are bonuses.

Having worked with soil my whole adult life, I've come to see it as a metaphor for so much else. This neighborhood and its inhabitants remind me of the abused farm fields I've taken on before—soil so mistreated that it felt and looked like hardened concrete. I nurtured those soils with compost and cover crops and mulches and watched as they recovered and came alive. After a few years you could reach your hands deep down into soil that had become loose and friable and fertile. Is it possible that here in this hardened landscape we could do the same, make useful and productive what had been abandoned and abused and forgotten?

Seann and I had no complicated vision for the corner of Hawks and Hastings, nothing that would require a degree in psychology or extensive experience in social service, just a reason for people to get out of bed each day, a choice to make between something that is life-giving and something that will bring you down.

Though uncomplicated, ours was a big idea, and for many reasons we knew that it might not work. I knew I could survive failure, but I worried about those whose lives have known nothing else.

Seann Dory (*left*) and Michael Ableman (*right*)

Halloween day, 2009. One hundred volunteers recruited through social media gathered at the Astoria Hotel's parking lot to clean up and haul away abandoned vehicles, bed frames, beer bottles, cigarette butts, shoes, clothing, used syringes, piles of trash, and construction debris, as well as to help build the wooden growing boxes for our first urban farm.

I remember standing in that parking lot on our first day of planting with four hundred 4-by-50-foot boxes full of soil waiting for the first seed or transplant. The transplants I'd brought to Sole Food were grown on my family farm. "Hardening off" transplants is a practice that normally involves gradually introducing tender plants to cold and sun, allowing

for the transition from protected greenhouse to open field. As we unloaded the plants from my van that day I had the thought that we ought to have piped the sound of sirens, rap music, and car horns into their protected rural greenhouse space before introducing them to this harsh urban landscape.

Among the crew of eleven people, not one of them had ever grown anything before. Yet they'd shown up and their hands were getting dirty. One was a man named Kenny, our very first hire. Kenny had worked with Seann at United We Can and jumped at the chance to help develop a neighborhood farm.

I came to this work with my own package of preconceptions and judgments. When I met Kenny, my first impression of him fit every stereotype about drug addicts and what they look like. Sporting a wispy, slightly graying goatee and wearing multiple chains around his neck, he was desperately thin and hollow-eyed, with a shaved head and a fast-talking skittishness that reeked of crack or speed.

I came to learn that, for someone who has been through hell and has had so much badass shit happen in his life, Kenny is a real softie inside.

When ladybugs show up on the produce while it is being washed and prepared for sale, Kenny will go to great lengths to save every last one from drowning.

People connect in many different ways. I don't need to be everyone's best buddy; sometimes I just like those relationships founded on mundane things, like a shared interest in cooking or food. I feel a special connection to those who have an eye for organization and an aesthetic that does not allow for things to be buried in disorder.

This is one of the things that I liked about Kenny from day one. He was the guy who noticed when the farm needed to be cleaned up, the tools organized, the fine details attended to. And while some farms worry about pesticide drift or safety around farm machinery, Kenny and the rest of us have different concerns. Residents of the Astoria drop used needles, crack pipes, condoms, and other paraphernalia out the windows, making work in the eight-foot stretch of the vegetable beds closest to the hotel a cause for caution. Kenny gets pissed off when the Astoria treats our farm as its dumping ground or when folks throw trash, and worse, from the windows or over the fence.

One of the roles that I have proudly accepted on every farm I've worked on has been head janitor. Most farms match people's visions— totally junked out with old equipment rusting on the edges of fields, hand tools left where they were last used, and piles of everything left everywhere simply because they might have some use at a later time.

On my family farm on Salt Spring Island, we have our "boneyard," but it's organized and managed so that when I need a two-by-four or a piece of rebar or a section of pipe I know where to find it. Visitors to my farm are always surprised when they see how neat and organized it is. "This is the cleanest farm we have ever seen!" they exclaim with some level of mistrust, as if a messy farm is some sign that everyone is too busy doing the real work of farming to put things away. "I don't have time to be disorganized or messy," I respond. I don't want to spend half an hour looking for a tool or repairing an implement that got left out in the rain. And I have an aesthetic that does not support junk piled everywhere.

At one of our year-end staff parties we presented Kenny with an apron that says THE ORIGINAL EAST VAN FARMER. Given his tenure at Sole Food, "original" works for Kenny in any context. He's been with us seven years, a long time for someone who has spent the last twenty years of his life strung out on heroin. I consider it a testament to our work that Sole Food is Kenny's longest-held job.

Unlike many social service projects, we have never seen it as our role to train people and move them on to other jobs. We've always wanted people to stay with the organization; we believe the farms—there are now four—and the work we do create safe zones, places to continuously return to. A job on one of our farms is one of the few meaningful engagements that our staff has, a place away from the hustle, the temptation, the noise, and the struggle.

None of us who've organized Sole Food really know that much about addiction, and so we don't diagnose or analyze or pretend that we are anything other than farmers providing meaningful work and a place to connect to. Kenny cannot turn to us for those things we cannot provide.

It might be that all we offer of real value is that rare constant, a touchstone, the stability that many of our staff have never experienced.

But going through the cycles of a year on a farm is also incredibly valuable. People who farm constantly see stuff die and other things come into life. When every day is spent getting down and dirty and close-up with those cycles, it gets into you, and you start to see the world differently, with a little more acceptance and an understanding that each of us is subject to those same forces.

Physically, Kenny is a walking miracle. He's been stabbed, held up at gunpoint, wanted by police; he's known most drugs. He's suffered bicycle accidents, illnesses, imprisonment. He's faced years of rehab. I am in awe of the life force that can keep someone going with that kind of hard-living history.

Yet when I talk to Kenny now, he tells me Sole Food has been a chance for him to achieve something—personal satisfaction, a place in the community: "It's a time when I'm happy," he says. "It gives me a sense of accomplishment." Sole Food has gotten some media attention, and Kenny, at least in his own mind, is a minor celebrity. As he speaks his hands are moving, he's fully animated, and his voice rises in pitch. "Everyone comes up to me and says, 'I've seen you on TV. What you're doing is a really good thing!'"

Kenny says he feels lucky, and proud, to be part of this farm. His work can turn a day around: "I come to work feeling miserable," he says, "and leave feeling relief and hope." Although my personal challenges are different, I can relate. There are so many times I too don't want to get out of bed, cold or rainy mornings when my back hurts and my hands are cracked from soil and water and I'm tired and curse the thought of having to get up and move through another harvest or day in the fields. Somehow I drag myself up, get dressed, and as soon as I am out the door and immersed in the open air, moving and responding to the myriad sounds and smells and sensations of farm life, I feel better, and I know that this is where I belong, and I feel thankful that I can be on the land.

Kenny tells me, "I've worked jobs where I've made a lot more money, but now I actually love my job, I love going to work. I still struggle, but this gives me an opportunity to help others." By Kenny's accounts, everyone who has stayed with us at Sole Food has gotten healthier. If you stretch your concept of what family is, move beyond the stereotype of Mom and Dad and the kids, you could say that the Sole Food farms and the community of farmers and eaters that rely on us are just that—a family. And for many of our staff, this family may be the only one they have ever had.

As employers—and we *are* employers—our goal is try to maintain that sense of family, even while balancing the expectations that employees will do the jobs they were hired to do. I won't say it isn't frustrating when, with crops ready to be harvested or new transplants waiting to be planted, a farmer misses his shift. But in guiding the farms, we accept that the lives of our employees are sometimes more chaotic and less secure than our own. So our employment model also allows for people to fall off the wagon and still keep a job. For Kenny that has been essential. When he is on, he's right there, 100 percent present and totally committed to the work and the team. But sometimes he still disappears into his opiate addiction or into rehab.

Kenny

Though Kenny and I connect in both roiling at disorder on the farm, Kenny has told me that he's had a hard time shouldering the kind of responsibilities we face at Sole Food, responsibilities that are inherent to farming. Over his whole life, he says, "I've gotten away with everything." Kenny works hard, but sometimes he doesn't show up. "When I miss work," he says "it's not, 'Why didn't you come to work?' It's 'Are you okay?'" Growing and selling produce is not the only measure of a job well done. This is a lesson I've taken from Kenny.

"I get to be in nature at the farms," he's told me, "and working with other people, and also be in the city. If it wasn't for my job I would be sitting in some basement not caring about anything. It's not about hurting yourself with drugs, it's about the damage you do to other people."

One of the wonderful and strange things that happen when you work with people on a regular basis is that your differences start to drop away. Farming together becomes a great equalizer. The traditional roles of "management" and "employee" are still there in subtle and not-so-subtle ways, but when there are so many bunches of radishes or chard or kale

to harvest and the sun is getting hot and the orders have to be delivered, you're all just part of the same farm crew.

———————

Planting tomatoes with Kenny that first year at the Astoria, everything was new, and a tomato plant was not just one more tomato plant, but something to behold. I am sure he didn't notice, but I watched him just looking at one of those plants, in awe of its fragile nature, and I sensed in that moment that he realized that plant was dependent on him for its survival. I saw his humanity that day, and I also saw how like that small plant, he too was a little fragile and dependent. And when the first red tomatoes ripened on those plants all of us came together to sample our work, and I got to watch that crew's expressions as their brains registered something truly amazing, something so new and so real and so beautiful.

Working this way, delighting in the world we share, is eye-opening. And still, with eyes open on this world—the pain, the poverty, the pull of drugs—it is impossible to not feel the suffering that is likewise common to us all. Rich or poor, sober or not, we all get our share. I've watched from a distance one of my family members go in and out of drug use. I've seen the misery it has brought to him, to my parents, and to all those who have gotten sucked into the powerful vortex that addiction can create. It has been painful keeping my distance from the suffering of someone so close, but I think it is easier for me to reach out to someone who is not.

I've told members of our crew at Sole Food, "You don't have a monopoly on suffering." I know that some of us got lucky, have access to more resources, were born into more privilege, got the right genes to be pretty or handsome, or have a more comfortable safety net to fall back on. And while it is near-impossible for me to live in this world and not feel the misery that so many people are going through, I know I need to recognize the difference between compassion and pity. I'm not sure the latter has ever provided much relief. Seann and I insist on compassion; we believe it drives all of us to some form of action—to make a donation, prepare a meal for someone, lend an ear, plant some seeds, or provide a job.

———————

It's a big day, our first large harvest since our little fledgling farm was planted, and Seann, Kenny, and I, together with the rest of our newly

hired crew, are gathering rainbow chard, lacinato kale, French Breakfast radish, and collard greens for the afternoon market held a few blocks away in front of the train station.

We'll soon be selling our harvest right there, to businesspeople fresh from their high-rise office buildings, travelers getting on or off trains or buses that stop a short distance from our stall, and residents from nearby apartments and condominiums. Sole Food gives *local* a whole new meaning, as the produce on our market tables is harvested within walking distance down the street.

On the farm we break off stems and leaves, pull roots, make bunches, and pile them into nearby boxes. It's a familiar rhythm for me, with familiar crops, and I realize there are probably hundreds of farmers at this very moment doing exactly the same thing—except accompanied by birds singing or wind rustling trees rather than the nonstop clamor of traffic and sirens and hustle.

We interrupt harvesting for one of our farm walks, a chance for me to share some techniques or a little philosophy, answer questions, and tell stories. And I realize that even as I am telling stories to make abundance real and visual for folks who may never have experienced it, I am feeling my own doubts and questions about what lies ahead. It feels odd for me standing in this parking lot on a street corner talking about soil microbes, optimal plant spacing, or the life cycle of an aphid. On my rural farm, not far from here, I'd be carrying on similar conversations, but there I'm mentoring young, well-scrubbed kids fresh out of college, most of whom have never known real hardship, all still hopeful and idealistic, too young for life to have slapped them around.

In most farming ventures that involve breaking new ground, the initial results can be fantastic, all that untapped soil fertility. We found that here as well. Kenny and the rest of the crew were making out well; our early plantings had thrived. The bunching greens produced huge leaves pulsing with vitality and color, radishes seemed to jump out of the ground, and tomatoes and peppers grew and flowered and produced with total abandon.

I knew our funders, our staff, and the broader community had been waiting for this—waiting, and watching, and expecting. It was overwhelming to consider that we were attempting to supply Vancouver's top restaurants and farmers markets with production quantities of food grown on a parking lot, with workers who had some significant challenges. Not to mention growing some hope and healing for our staff and the broader community in the process.

So it was a relief to witness such fecundity and productivity and blatant health in the midst of the struggles of this neighborhood. I knew this first great crop was important for our crew, too. They needed to see something thriving and working. I suspect this is why many of those who were on that first farm crew, people who may not have kept a job for very long—people like Kenny—are still employed with us after all these years. Early success was essential. And we got it.

CHAPTER 2

The Unpaving of Paradise

From the farm, walk across Hastings and head down Hawks and within a block and a half, the hustle and shuffle slows and the street scene is dominated by newly painted houses, gardens, bike lanes, and cafés. This neighborhood, the oldest in Vancouver, was scheduled for demolition in the 1950s, but residents banded together and protected it. Because of its proximity to its less reputable neighbor, the Downtown Eastside, Strathcona's real estate prices lagged behind those of the rest of the city. Now it is having one serious late-life makeover. Run-down houses that were languishing on the market just a few years ago are today selling for well over a million dollars.

Seann and I sometimes walk to the Wilder Snail, a little corner store and café in Strathcona, to take our lunch break, find some quiet, or dry out from a soggy Vancouver day. Everything in this café seems so clean relative to the nearby farm and neighborhood, and the warm colors, good lighting, and hot drinks provide a brief respite on those days when the wet and the cold have found their way through every layer of clothing.

We've known this spot for years. It was over breakfast burritos and soy lattes at the Wilder Snail that we first dug into the details of our grand urban farming enterprise. Everyone at this café is plugged into some kind of electronic device, so we fit in well tapping out our visions on laptop keyboards. The coffee shop has always been a comfortable place for us to haggle over ideas, to dream, and those bits and bytes of data we put down then remind me now of the thousands of tiny seeds we've planted early each spring, so small and innocent when placed into flats, eventually growing and demanding energy and time and space.

The farm at the Astoria was the proving grounds where we could experiment and develop our systems, figure out how to grow in containers, and determine what crops we could make work agriculturally and financially on intensive spaces in the city. That first harvest day yielded two tall tubs of spinach and salad, a box of arugula, three cases of rainbow chard and three of lacinato kale, one case of French Breakfast radish, and one of Cherriette radish. It was a humble harvest, but the excitement, pride, and hope for the future it generated were worth far more than the $750 in cash it yielded.

The Astoria farm also provided experience with creating an employment model that could help those we were reaching out to while maintaining good farm production. We hired eleven people that first season, we paid out $60,000 in salaries, and we generated over ten thousand pounds of fresh product with a value of close to $70,000. But for me, this site and these numbers were always just a starting point, and the throngs of down-and-out people who walked by that farm every day were always a reminder that there is so much more that we could do, so many more people we could train and employ, so many more tons of food we could produce.

From the beginning Seann and I hoped we could grow the project to multiple sites throughout Vancouver. The original plan we wrote, huddled

over our laptops at the Wilder Snail, had us expanding from the half-acre farm at the Astoria to thirteen sites and fifty staff within the first five years. To achieve this, we figured, would require our raising a million dollars. We'd also have to convince folks who had never held a long-term job or had a permanent address that two white guys not from around here were more than just romantic do-gooders, that we could provide concrete results in a community that had seen numerous schemes come and go. We'd have to figure out how to safely grow food on pavement and contaminated land. And we'd have to sweet-talk local bureaucrats to stretch and bend municipal codes to allow all this to happen.

We didn't know it then, but Sole Food was up against a municipal bureaucracy that was comfortable with building codes, residential and retail uses, bike lanes, parks, libraries and schools, roads, public transportation and garbage collection, even community gardens, but not with real farming. And in the coming years we were all in for an education in how to navigate a regulatory system that did not understand or even recognize growing food commercially as an urban activity. If asked, most people would say that farms and farmers markets were important to the civic health of our cities, even as much as our schools, churches, museums, and sports facilities, but translating that ideal urban life into a working reality was not so simple.

Greg

In those early days, we were operating under the auspices of the Downtown Eastside charity United We Can. They had experience with the drug- and alcohol-addicted, the "hard to employ," as these people have been euphemistically identified. The sidewalk in front of their bottle-recycling depot on Hastings Street, the largest in Vancouver, was a central hangout spot for men and women from the neighborhood, a vibrant social scene where, for a price, almost anything could be obtained.

United We Can provided a nonprofit umbrella for our early work at the Hastings Street farm, and we stayed with them until December 2011. Almost from the beginning of our first farm, as our plans to expand became clear and possible, we realized we needed to form our own charitable organization. But it took time to sort out how to do that and

Weed Wisdom

Our goal is always to cultivate and never to weed. And while it is true that there is some subtlety in how the terminology is used or interpreted, it is an important distinction.

Cultivation aerates the soil and controls weeds at an early stage, before they become established. Cultivation takes care of weed pressure often before the weeds are even visible. Weeding is what you must do when you are too late to cultivate. The energy required to cultivate is a fraction of that involved in dealing with weeds after they have increased in size and gotten a foothold. Hitting the right timing requires constant observation; a day late can be a half day too late.

Using the right cultivation tool for the conditions makes the job more efficient, requires less energy, and is more comfortable. There are numerous versions of the traditional hoe; each one has a particular application. Experiment with different tools and find the right one for your body, your soils, local conditions, and particular crops and stages of crop development. We use a collinear hoe, which is simply a thin blade attached to a handle used in a motion similar to shaving or sweeping—an almost effortless glide—just below the soil surface. When it's done well you can cover large areas quickly.

And while we attempt to provide our vegetable plantings with as little competition as possible, we also have to acknowledge that a farm is much more than straight rows of single species made up only of what we put there. While production demands require that we always give our planted crops the upper hand, we need to recognize that those plants that volunteer in our cropping system also have a role. What is the appropriate balance between the cultivated and the wild? Are they weeds, or plants out of place? Consider that:

- All cultivated plants were once weeds, selected from the wild for nutritional or culinary characteristics.
- The weeds we remove from our fields are often more nutritious than the cultivated plants we leave behind to harvest.
- Weeds are windows into our soil. Learn to read them.
- Weeds provide nectar for bees and habitat and food for wildlife.
- Weeds often appear to fill an ecological gap, to heal and to balance our soils.
- Weeds provide an essential reservoir of genetic material.
- Weeds act as living mulch, and nature's cover crop, protecting soils from wind, sun, and water.

where to find the financial help. To support our big ideas we needed some big money, and although I had successfully raised funds for projects before, I dreaded having to go out begging again.

In 1987 I was given one year to raise a million dollars to preserve the farm in California that I had been operating but did not own. The land was zoned for fifty-two condominiums, a fate I could not accept for a farm that was so loved by the community. We raised those funds in eight months. That experience was both empowering and exhausting. But it taught me that asking for money from people who had plenty of it was ultimately less about begging and more about providing them with an opportunity.

Even as we planned the Astoria site and throughout 2009 its beginnings as a working farm, I knew that if the project (and urban agriculture in general) was to be taken seriously; if it was to be more than an idea, a small garden, or a dream; if it was going to be truly agricultural; then it should operate like every other farm and generate its income by the pound. As a farmer I was never comfortable with the idea that Sole Food would have our hands out forever, in part because of that deep sense of self-sufficiency that comes with the profession and in part because of my own pride.

In the United States $46 million a day of taxpayers' money was being spent to subsidize large-scale industrial farming. I have read the investigative reports about fat subsidy checks being mailed to farm owners with addresses in Beverly Hills and Pacific Palisades. Even had I qualified for those subsidies while farming in California, I wanted nothing to do with them. Since the mid-1970s I had been operating small farms as successful businesses, and I felt this one could be as well. So Seann and I wrote a plan that projected financial stability within five years.

The thirteen sites we planned for would be a minimum of half an acre each, with a manager responsible for every two sites, and a core staff in numbers depending on the scale of each farm. Each neighborhood farm location would be treated like a separate field in a larger contiguous farm, and we would rotate crops throughout the separate sites just as we might on a larger farm. There would be one central location that would receive the products from all of the farms. All of the washing, processing, packing, and distribution of those products would be carried out from that site. And there were plans for a composting site that would serve all the farms.

Our original plan and budget even had one staff member with professional counseling experience, who would move around the various sites meeting with our farmers. We hoped this person would help to navigate through the ups and downs we watched our farmers go through, improving their experience and allowing the farms to operate with more stability.

I have since had to accept that the social piece in our farming, employing people with barriers to traditional employment, will probably never allow us to compete and operate like any other farm. For instance, during our first harvest I watched Freddy, whom we patiently nurtured and trained as a farmworker, struggle to participate successfully. He had no control over his grip, and by the time Freddy managed to place a bunch of arugula or radishes into the harvest box, he had squeezed it to a pulp.

It might have been simpler to hire skilled farmworkers. Had this been some other farm, I might have considered doing just that. But Sole Food was never intended to be just another farm. Its unique mission is all about the people we employ.

I quickly learned that it could take two to three times longer for our crew to transplant chard or harvest a pound of salad or cultivate a bed of spinach than it might on another of the farms I've worked on. And while I tried to demonstrate how to do farm tasks more efficiently—pre-marking and dropping plants in advance of transplanting, placing the harvest box higher off the ground and closer to where someone was harvesting, cultivating with tools that are sharp and when the weeds are small—I had to accept there was a major added cost to our employment model.

It wasn't immediately obvious to everyone involved—Seann and I included, some days—but Sole Food, from the very beginning, offered huge benefits to the men and women we worked with and to society as a whole. Several years into the project, a team from the MBA program at Queen's University conducted research on our farming enterprise and determined that for every dollar Sole Food spent on employing people who are "hard to employ" there was a $1.70 combined savings to the prison and legal system, the health care system, the social assistance networks, and the environment through carbon sequestration and the energy and transportation benefits that our über-local farming system provided.

We were thrilled to get that kind of hard data, real proof that our economic impact was far greater than the bottom line on our quarterly reports. But even before this good news arrived, we knew that studies and reports about social change would not translate into actual dollars. The shortfall in our budget, which we'd anticipated in our earliest models, had to come from somewhere. Two years into the project we estimated that our employment scheme and our broader social goals would continue costing us approximately $250,000-plus more than our annual budget, and we accepted that we'd have to raise those funds every year.

And from the beginning, and with each new farm we planned, we had to face the reality that you can't just pop a hole in the pavement and plant a tomato or put seeds into ground that once housed a gas station, a warehouse, or an industrial facility. Knowing that all of our sites, and in fact the majority of existing and potential urban farming sites in the world, either were paved or had native soils that were too contaminated to grow in, we had to come up with a container system that isolated the growing medium from contamination and was movable, durable, and affordable. We'd first done this with planters at the Astoria site—using dimensional lumber, two-by-eight fir planks, and four-by-four posts. And while we have since designed and had manufactured some much-improved longer-lasting plastic models, we are still working to find a way to produce those containers at a cost per unit that could be absorbed into any start-up urban farming enterprise.

The numbers did not lie. Seann and I knew it. Scaling up, growing tons of food and dozens of jobs, establishing production-scale urban agriculture, would necessitate major capital investment, specifically for boxes and soil, that would never be required in more traditional open-field agricultural systems.

Water Wisdom

It does not matter where in the world you choose to farm; if you are growing fruits or vegetables, chances are you will have to irrigate. Regional climate, soil conditions, crop choices, and water availability will influence this practice, but there are some universal truths about water and how we use it.

Farming in California during the drought in the mid-1980s was a great education. The local water district put forward restrictions based on historical use. Because we had already been conserving water prior to the drought, we were punished with a smaller allotment, while those who had previously been sloshing it around were rewarded. But the severe water restriction turned out to be a blessing. It forced me to learn what and how I could grow with very little water.

We adjusted our crop mix so we could focus on less thirsty crops. We researched varieties that lent themselves to drier conditions, and we changed our planting and growing systems to accommodate dry conditions. Tomatoes were planted deeply and farmed without any water, which resulted in fruit that was intensely flavorful. Melons and beans were treated the same. We obtained vast amounts of spent horse-stable bedding that was being dumped into the local landfill and used it as mulch. We improved our water application methods by using exclusively low-volume drip, reduced the frequency of our watering, and became ever more precise with our irrigation timing.

As our soils improved and the organic-matter content of them increased, so did their ability to hold moisture. Poor

And what, we had to consider, could possibly entice the owner of a parking lot or a future building site to allow a bunch of drug addicts from skid row to move in and start growing food? Gaining access to valuable urban land required that we find a way to appeal to the needs of both private and public landowners, the developers and municipality in equal measure. No matter how we did the math, our business plans made plain that we would never be able to offer enough money. And most urban landowners do not want farms and gardens on future development land because of the public relations ramifications of having to ask them to leave. The value of the land was too high, we had to offer something other than hard cash, and we needed to demonstrate that we would be good neighbors.

soils can act like sieves or can be so locked up they do not drain. We wanted soils like sponges, soils that absorb moisture and then gradually release.

Open fields have that giant reservoir of moisture that buffers dramatic wet and dry cycles. Containers tend to dry out faster and can have soggy bottoms. Irrigating in containers generally requires more frequency and more careful monitoring. Raised boxes can overheat in warm weather, and irrigating the containers also functions as a cooling method.

Remember these tips:

- While hydrologic systems do replenish themselves, water is a finite resource and should be treated as such.
- Seventy percent of the world's fresh water goes to irrigated agriculture. Due to inefficient application and transport methods, only a fraction of that water is actually utilized by the intended plants or animals.
- Not all water is created equal. Water quality varies dramatically and must be considered when irrigating. Mineral and salt content, temperature, and pollutants all have an impact on soils and on plants and animals.
- Creative use of irrigation can dramatically affect food flavor, texture, and size.
- Application technologies (such as drip tape), soil improvements, and mulching can reduce water use, improve food quality, and diminish weed pressure.
- Providing the conditions for rain and irrigation water to slow, spread, and soak in rather than run off is one of our most important responsibilities as farmers.

In 1999 Rudolph Giuliani, then mayor of New York, threatened to auction off 112 community gardens to the highest bidder. Taking on demonstrators dressed as ladybugs, tomatoes, and earthworms became one of Giuliani's biggest public relations nightmares. I added to the uproar by bringing together some well-known friends—Alice Waters, David Brower, Pete Seeger, and several others—who converged on New York City for a series of events sponsored by a local organization called Green Guerillas. At the eleventh hour, Bette Midler and the Trust for Public Land came through with the money to save those gardens. But in fact, that land was incredibly valuable. And it is no easy task to compare the value of fresh food or gathering places for neighbors, or the power of community, with the millions of dollars that were

at stake then in New York City, and the millions that remain up for grabs in Vancouver today.

———————

I've often thought someone should develop a Reality Check app for smartphones. It would make the phone flash and buzz and chirp each time its user comes up with a big idea, and then, through some powerful algorithm, issue a series of practical warnings detailing the associated yet unforeseen challenges, stresses, and real costs that lie ahead.

By the winter of 2011 Seann's and my original plan for thirteen sites in five years had shrunk to a more reasonable five sites over three years. Finding employees to train and food to grow had not been a problem at the Astoria site, and we didn't anticipate significant troubles in those terms at future sites either. But we would need more land. And to access additional land in Vancouver we had figured out we needed to create a lease arrangement that was attractive to landowners. Our arrangement, we knew, would have to appeal to their needs and instincts; it must promise public relations benefits for them, beautify their sites, provide some tangible tax or financial benefit, and demonstrate that we could vacate on relatively short notice. There was a long list of things to do, a tall mountain to climb, and we needed a map just to get to the trailhead.

If there is one thing I've come to know throughout my career, it's that one essential element in manifesting a vision in the world is the powerful act of clearly writing down one's goals and intentions in great detail. Stating one's intention is not some psycho-spiritual woo-woo concept, but a very real way to move from idea to manifestation. Identifying a specific goal forces you to clarify where you are going and how you want to get there. Writing it down situates an idea outside your mind and into the world. With a new level of clarity, and with work and projects that are well grounded and go beyond personal benefit, you also open up possibilities for help to arrive in many ways and from many directions. This has been my experience over and over.

My practice of establishing goals, and writing them down, is itself grounded in a quotation I often return to, one typically attributed to Goethe despite scant evidence it actually came from him. Still, I love the line: "Whatever you can do or dream you can, begin it. Boldness has genius, power, and magic in it." Our kind of boldness—as with most kinds—also had daily work and persistence in it. We had already begun. And so Seann and I had to continue on through tedious meetings, drafting spreadsheets and business models in the local café, developing goals, realistic concepts, and a phased approach that would eventually be the road map for Sole Food.

Over months of such meetings, we discovered that the unpaving of paradise was not going to happen overnight. Peeling back the layers to create production farms in the city was actually going to require a collaboration of mammoth proportions—beyond what United We Can could offer us, beyond what we could ask of them. And so we wrote and rewrote our vision, continued to improve our farming systems at the Astoria, and began seeking out the support that we would require to manifest each layer of our plan.

The description of our new plan opened with a line from Masanobu Fukuoka: "The ultimate goal of farming is not the growing of crops, but the cultivation of human beings." The document provided an overview of our social and agricultural goals, along with a description of the potential market for our products, whom we were planning to employ, and the principles of that employment. It described how our project would benefit the food insecurity that plagued the neighborhood, our

Sage

community engagement strategies, and a work plan and time line that provided the details of how the project would be phased over three to five years. And it included biographies of the principal players, budgets and income projections, capital requirements, an organizational chart that clarified the management and staffing structure, an operational chart that revealed the flow of the project from seed to table, and some abbreviated case studies of comparable urban production agriculture in Cuba and projects I had been involved with in the United States.

As a farmer I sometimes bristle at the use of the word *agriculture* to describe much of what people now call urban agriculture. I know it's just words, but what else do we have to communicate? Most of what is being called urban agriculture is actually urban *horticulture*. No judgment intended here at all; we need the garden-scale projects as well as the larger-scale ones.

But calling a garden with a few boxes of vegetables "agricultural" is like referring to yourself as a mechanic because you change your own oil or an actor because you once performed in the school play. It is important for farmers that the world begin to understand that doing agriculture well requires a very sophisticated and complex set of skills. These skills take years to develop, and they require a deep understanding of soils, insects, biology, botany, mechanics, physics, marketing, labor management, and on and on.

Experienced farmers come to know that observation and careful timing of all farm activities is critical. Cultivating when weeds have just emerged, harvesting beans or greens when they are at their optimal flavor and texture, irrigating at just the right moment to enhance food quality and save water and time. Harvesting well, for example, has so many subtleties and particularities. All bunches—carrots, beets, or radishes, say—are not created equal. A well-made one has roots that are matched in length and girth, are positioned at the same level, and have had their cosmetically imperfect leaves removed.

Every art or craft has its own unique elements recognized and appreciated only by its longtime practitioners. I have come to appreciate these elements in the farmwork of others. Some are subtle, some are more obvious. I look for the way a crop has been cultivated, the feel and smell of the soil, the quality of the farm products, or the speed with which a farmer can move down the rows of a field making perfect bunches. Doing so requires a deftness of hand and a good eye.

As a mentor and employer, I've found it remarkable that you can take ten people into a field of vegetables, carefully and painstakingly show them all how to harvest, demonstrate the speed and efficiency required to make it pay, and each person will come up with his or her own interpretation and level of quality. Every so often, though, someone gets it right off. Donna was one of them.

I'm not sure why, maybe good hand–eye coordination developed from years of rolling joints, maybe just a natural appreciation for detail, but from the first harvest at the original farm on Hastings Street, Donna could crank out the most beautiful bunches of radishes.

Like many of our employees, she came to us from the neighborhood that surrounded the farm. She was living just off Broadway and Fraser in a building that used to be social housing but got converted to condominiums. She was forty-two years old when she came to work with us. In my first encounter with her she was hesitant, standing back to see who I was, what I was doing here, and whether I could be trusted.

Her hair was pulled back in a ponytail with gray-streaked bangs hanging down into her face. She had the pallor of someone who's been smoking for a very long time. As I worked alongside Donna, I came to learn a few things about her life before Sole Food. Sexually abused by her father as a child, Donna lived in fear her whole childhood. She watched as her mother used intravenous drugs—"there were a lot of people sitting around sticking needles in their arms," she told me—and though Donna swore she'd never do anything like that, at age ten she began smoking cigarettes and pot. "I was often by myself," she told me, "and I saw this joint in the ashtray and picked it up."

Over three and a half years the family moved fourteen times. At twelve, she told her mother and brothers she'd been molested by her father, who was by then married to another woman. When she told her stepmother that her father was a pedophile, the woman denied it.

The years that followed were a spiral. After Donna's mother died and her primary relationship was lost, she was broken. She was sick in bed for three or four months after that. To make money, Donna would play her guitar at the neighborhood liquor store. She made friends.

"I knew a lot of criminal-type marginalized individuals," Donna has explained. "Everyone was smoking crack. I remember walking past this bookstore. Every day after hours a bunch of friends would be in there smoking crack. One day I went in. I thought to myself if I get really messed up I can just leave or go away. I didn't understand the full nature of addiction."

Crack became her drug of choice, and her use increased beyond her control. By that point she was raising four kids, but never making much money. She did what she could to keep food in the house.

"But in the end I just didn't care anymore. I said, 'Fuck it. I'm getting high.'" Which she did for about a year. Donna figured the older kids could look after her youngest, Kegan. But then she got sick with an antibiotic-resistant infection. She shrank to seventy-five pounds.

Everyone, she told me, hits bottom: "This was mine."

Donna found help; she entered recovery. But that first year was tough, she told me. She worked for a time as the ladies' bathroom attendant at the Carnegie Community Centre, an outreach facility, but was fired because her employers thought she was using again.

Out of work again, Donna started volunteering at the Downtown Eastside Neighbourhood House, not far from our Astoria farm. "And then," she told me, "I saw Sole Food pop up on the corner."

Donna was not the easiest person to get close to; she could be guarded, protecting herself in subtle and sometimes not-so-subtle ways. But I could see right off that she had the intelligence and the eye to be an excellent farmer. And she had the desire. It is a wonderful thing for me—and her—to recognize small successes, see beyond the struggle, the hardship, the trauma.

THE UNPAVING OF PARADISE

I wasn't present for her original job interview, but she later told me that she had wanted the job so badly that she wore a shirt that had flowers and vines in the design because she thought it would help her chances. She may have been right.

Donna worked with us for four years, during which time we expanded Sole Food from one to multiple sites. The expansion would prove stressful for everyone, especially those who had come to work with us at the beginning, when we were focused on a small crew and the purity of our vision was still intact. This was true for Donna—she had trouble riding through the growing pains we experienced as an organization. Growing the enterprise made it difficult for Seann and me to give the same level of attention to the longtime employees as we had in the beginning, and I believe this contributed to Donna's ultimately leaving.

Donna arrived at Sole Food with the smarts and a strong desire to learn. She came to possess the necessary skills. And she did so with pride. "What's the use of being sober if you don't do anything with your sobriety?" she once said, describing her work at Sole Food.

Donna

Our cities have evolved on the idea that food comes from somewhere else, that farms are far from cities, and that urban dwellers are like baby birds in the nest helplessly waiting to be fed. Today, however, we can see the beginnings of urban planning and development in North American cities addressing food production as an essential element of society. And while a new awareness is forcing a shift in thinking among planners and policy makers, most of that thinking is still focused on garden-scale projects and not on agricultural production and the jobs it can generate. As I see it, any project that's essential to our cities must draw on those people who call those places home. From our earliest days of planning our expansion, sitting together at the Wilder Snail, Seann and I had Donna and the rest of our small team in mind. We were thinking agriculture, not horticulture. To generate more jobs and more and more food throughout Vancouver was a goal shared by many people in the city when we were getting started. To do so we would need lots more money.

On March 7, 2011, I was copied on an email to my eldest son, Aaron, from Jessica Lubell, a young woman who had picked him up hitchhiking years before.

In the email, Jessica explained she had been at dinner the other night when discussion turned to sustainable farming. A man she was dining with brought up my 2005 book *Fields of Plenty*, which had been inspirational in his own agricultural endeavors, he told her. When she mentioned that she knew Aaron, and described the twist of fate that had brought them together, the man immediately asked if she would be able to coordinate an introduction.

"The gentleman," she wrote Aaron, "is Frank Giustra, who, with a mere click of your mouse, assisted by the ubiquitous Google, you will see is quite an extraordinary man. Frank is fascinated by organic farming and I believe your father's book was pivotal in his interest."

I was not familiar with Frank, so I did what most people do these days when they want information, the very thing Jessica had suggested: I stepped into the cyber-abyss and checked him out. Frank is a billionaire, I discovered, who'd made his fortune in financial services, mining, and Lionsgate Films, which he founded in 1998. In 2007 he pledged $100 million to former president Bill Clinton to start the Clinton Giustra Enterprise Partnership, a project of the William J. Clinton Foundation. Here was someone I needed to meet.

So on a rainy day in April I made my way from the gritty world of East Hastings Street to the Bentall building in downtown Vancouver, where I boarded the elevator to the penthouse floor.

Over the years my work has put me in contact with various wealthy or famous people. I am occasionally impressed by their outer accomplishments but not always by how they carry themselves personally in the world. Frank seemed different. I liked him immediately. He greeted me with a warm and confident smile and a firm handshake. Frank was casually but impeccably dressed in a white shirt without a tie, black pants, and some very classy shoes. I had made an extra effort to put on some nicer clothes for this meeting, but I'm not sure I got it right. Surrounded by so much elegance and style, my shoes in particular—the leather cracked and covered in dirt, mismatched laces—made me a little self-conscious.

I sat down across from Frank in his office overlooking the city, and—oblivious to my shoes—he expressed sincere interest in who I am and what I do. Despite all his success, his energy was low key, and I did not sense that high-strung arrogance all too common among the famous and super rich. As I looked around I saw photographs of him with Nelson Mandela, Bill Clinton, and others of that caliber, people I admire.

That morning, sitting on the couch in his office drinking tea, we had a productive conversation. He told me about some projects he was working on—including a farm he had purchased and a still-undeveloped business concept for bringing Old World food traditions to the masses—and I described my work with Sole Food Street Farms. In a few words, I laid out the goals Seann and I had set for ourselves: multiple farms, expanded food production, employees from the Downtown Eastside numbering in the dozens. Frank said he was interested in my help with his agricultural and food endeavors. With a boldness that I hoped contained some magic, I in turn told Frank that Sole Food was looking for an angel to give us a million dollars to support our broader vision and expansion. Frank agreed to visit me at the Astoria in the near future to see the farm.

Anton

Two months later Frank and several people who worked with him met Seann and me at the Astoria farm site. It was an incongruent scene—suits and ties and polished shoes gingerly making their way through the soggy pathways of the farm surrounded by the smells, sounds, and life of this underbelly of the city. As we often did, Seann and I each explained the social goals, the unique aspects of the intensive agriculture, the challenges, and the desire to take this little model and expand it to more sites around the city.

Frank immediately took to the idea and pledged on the spot to be our "angel." In turn I agreed to advise him in developing the farm he had purchased, and to consult with him as he expanded his broader business ideas within the food and agriculture movement.

Standing in the parking lot with Frank and Seann,

looking over the farm we had so lovingly and laboriously developed, I felt a sudden jolt. With Frank's decision to help, Sole Food was more real than it had been just minutes before. Our grand idea had now germinated. This was the moment that eventually does happen when you believe in something and never give up. And yet along with the excitement I remembered to be careful what I wished for. Each moment like this, and I've had my share, also comes with the knowledge that once an idea has been planted and it comes to life, it must be nurtured, protected, and cared for.

The seed that had been planted in fertile ground had been swelling, waiting for the right conditions to emerge. That day in my mind's eye I saw the first tentative fragile green shoot emerge.

CHAPTER 3

The Urgency
of Spring

Urban farming is increasingly capturing the public imagination. Fruits, vegetables, even grains, dairy products, and meats are being produced on derelict land, parking lots, street corners, and rooftops—spaces that have slipped through the cracks of the high-value real estate markets of many cities. Though too often city dwellers still view farming as something that someone else, somewhere else, does for them, they now have opportunities to participate in the amazing and dynamic world of agriculture. In some regions these ideas have even captured the imagination of mayors, councilmen, and county supervisors. But the reality of what we were proposing to do with Sole Food and the scale on which we planned to do it were unique.

On the leadership side of city government in Vancouver, we had the perfect storm. Our timing for the start-up of this venture coincided with a mayor and city council that, under the flag of the Vision Vancouver initiative, had decided that urban agriculture was a key goal in making ours one of the greenest cities in the world.

Once an organic farmer himself, Mayor Gregor Robertson was the founder, with his then wife, Amy, of Happy Planet juice company. They established the juice business to process the seconds produce from Glen Valley farm in Mount Lehman, British Columbia, which they started in 1991. I first met Gregor at a meeting of food and farming activists in 2001. Once while staying at the mayor's house in Vancouver I watched him return home on his bike and proceed to empty onto the kitchen table a canvas bag full of wild edible mushrooms that he had just harvested. My kind of mayor. Gregor understands our work, and his leadership combined with our plans and skill set aligned at just the right moment.

Soon after he was elected in 2008, Gregor brought in Sadhu Johnston from Chicago, formerly that city's "green czar," to fulfill the same role in Vancouver. As deputy city manager, now city manager, Sadhu became a staunch ally of ours, even at those times when we clashed with the city bureaucrats who worked under his direction.

These upper reaches of Vancouver city government understood what we were trying to do and saw us as a way to fulfill a keystone piece of the city's goals. While many world cities have the desire to become the greenest, Vancouver is one of the only cities in the world to create a detailed plan. Its Greenest City Action Plan (a document produced by Vision Vancouver) has ambitious goals and time lines to reduce greenhouse gas emissions, implement transportation alternatives, become zero-waste, establish renewable energy strategies, and create more local food initiatives. I'm not sure who could be the judge of which city in the world becomes the greenest, but Vancouver is a great benchmark to encourage competition. Still, high-level alignment and support does not always trickle down into complex bureaucratic municipal systems that were established to regulate conventional infrastructure such as the construction of a garage or a school, the remodeling of a kitchen, or the building of bridges and roads.

In fact, from the earliest days on our Astoria farm and especially as we began to expand to other sites, it became clear that our needs were entirely foreign to the existing system, totally different from anything that had ever been done in the city. Building inspectors, for example, did not differentiate between a brick-and-mortar building designed to house auto parts and a tunnel house used for extending the growing season, which is merely a sheet of six-mil plastic stretched over a steel frame. And this was just the start. We soon discovered that there simply were no municipal codes that addressed greenhouses, or composting, or multi-acre parking lots full of food.

This *should*, but does not, surprise us. After all, much of agriculture originated in and around what are now large cities. The Sumerians in 5000 BCE established sophisticated irrigated agriculture in and around some of the world's earliest cities in what is now southern Iraq. In Mexico City the Aztecs created chinampas along the canals, floating food gardens that provided a significant amount of food to that city. And in the fifteenth century the intensive agriculture in Machu Picchu in Peru provided almost complete food self-sufficiency.

And even as farms have been pushed to the outskirts in modern times, separated from cities by suburbs, certain urban areas have accommodated agricultural enterprises with great success. For example, Paris during the nineteenth century converted the "waste" from its predominantly horse-drawn transportation system into thriving French intensive farms. Some estimates suggest that these highly productive urban farms, which used systems that many organic farmers unknowingly replicate today, made up close to one-sixth of the entire city of Paris.

Most contemporary feedlots and slaughterhouses are located in rural areas, but in the Midwest of the United States in the late 1800s, feedlots and slaughterhouses were common in cities such as Chicago. Animals ready for slaughter were brought into urban stockyards for processing, and thousands of urban people were employed in that work. And long before the mega dairies of today, massive dairy operations existed in New York City into the mid-1900s, feeding cows on the spent mash from the multiple distilleries and breweries that operated there at that time.

More recently, when Cuba's reliance on Soviet food, seeds, and fertilizers ended almost overnight, it developed one of the best models of urban agriculture in the world in and around Havana. Like the Jews who developed extensive food production in the ghettos of Europe in the

1930s, or those who have no other place to grow food than on the garbage dumps in eastern Kolkata, India, these efforts are all about survival.
The urban farms I saw in Cuba incorporated some of the most innovative systems I have seen anywhere in the world—mycorrhizal root dips,
magnetized water systems, and a level of food crop diversity and productivity that was astounding. But these innovations and creations evolved
out of the simple fact that people were hungry and needed to eat.

Today most municipalities have some experience with community
gardens, and many have even put language on the books to address such
land use. Production farming, however, is not so common. With Sole
Food, we've always worked under the assumption that to make urban
agriculture truly *agricultural*, we would have to grow serious volumes of
food on acres of land and create the jobs to do that. There was little
precedent for that scale and none of the required infrastructure.

No one knew this better than the miracle worker Kira Gerwing, a senior
planner with the city of Vancouver. A roll-up-her-sleeves, no-bullshit,
hard-driving, get-it-done individual, for several years Kira became our
go-to person in the city. She was our gal embedded in the system.

I first met Kira in early 2010. Seann and I were in the midst of hatch-
ing our plan to expand the project. We had just secured the funding to
do so, and we needed inside support to grease the wheels of city government. Kira was a nine-year veteran representing the Downtown Eastside.
She had experience and knowledge of the planning process, and most
important she knew how to navigate the bureaucracy. She married those
skills to a deep commitment to the special needs of the folks who lived
in our low-income neighborhood and a real belief in the work that Seann
and I were doing.

Every problem, every bureaucratic hurdle, all of the endless challenges of our incongruent urban farm project went to Kira. She
negotiated the development permit with the city for the Astoria site to
allow us to farm there, helped secure leases and permits on three other
city-owned sites, and helped us obtain a grant from the city for our supportive employment work. When, at one point, all hell broke loose with
the Department of the Environment, Kira worked behind the scenes to
resolve that crisis.

While I am sure that we drove her crazy, I also know that she shared
our ambition to do something that had never been done before. Along
with the innumerable day-to-day challenges of managing the Astoria
site and our staff, all through 2012 with Kira's help Seann quietly slugged
away at the permit process,. I was happy he took on this part of the work;
I don't have much patience with dealing with bureaucracies, and I

needed to stay focused on the agricultural and financial needs of the project and its expansion.

Part of what fueled Kira, and kept her working on our behalf for as long as she did, was that she saw Sole Food as the poster child for the renewal at the heart of Vision Vancouver—the notion, she once explained to me, of "revitalization without displacement." Kira believed that if Vancouver was to meet the goals established by its 2011 Greenest City Action Plan, and if the whole of the city was to come alive and be economically and socially healthy, Sole Food would be a key player. "We knew where we wanted to get to," she once told me. "Urban agriculture was the top priority of the Greenest City goals. It hit on the food side, the green jobs and green economy side, and there were other higher-level city frameworks that it aligned with like our Healthy Cities program." Sole Food was also based in a part of the city other development projects would sometimes ignore; but we, Kira included, believed in pursuing economic transformation in ways that included disenfranchised and low-income people.

Kira worked with us until her departure from her job with the city in fall 2012. And through those early days, her work on behalf of Sole Food was relentless and risky. "I needed to be a diehard," she's explained. "I had the ability to go to the hierarchy to actually do something about it. I would go to the director of development services or the chief building official when either of their staff was being obstructionist. I would camp out in front of their offices until they would return from a meeting and I'd be sitting on the floor waiting. I'd catch them in the two to three minutes they had between meetings so I could point out the idiocy of the day." This worked for us. After dealing with Kira's insistence, often those officials would just make a call and move the ball forward. A permit would be signed. Permissions were granted.

Kira once described city building to me by first explaining what it is not. It is not like baking a cake. It is not like taking a rocket ship to the moon. Those tasks are at once both simple and complicated, and they have replicable solutions. "City building," she said, "is more like raising a child. Sometimes I think kids only see black and white and only what is on the page. But usually there is no black and there is no white and there is no page." With cities, as with kids, she said, "You have to embrace that complexity."

Embracing complexity, we were soon to realize, was a theme that everyone involved with Sole Food would have to get comfortable with.

Responding to input from our funders, Seann and I reduced our original plan for thirteen farms to five. At first this was a difficult decision for us to make, but in hindsight it was essential. Settling on potential new farm locations was challenging. We had already faced the difficult experience of several time-consuming and expensive false starts in the city, so we were cautious and careful.

Our false starts included a large rooftop prospect, which was halted due to insurance costs, and a beautiful parcel in Chinatown, which due to local politics became unavailable. Later the city offered a contaminated hillside next to an overpass, again on Hastings Street. Neither Seann nor I was involved with that offering, which was accepted by board members of United We Can. As is almost always the case, when people who know nothing about land or farming make decisions about land or farming, things do not work out well.

When I first visited that site, I knew it was a bad idea.

Over the years I've learned to read land, and this land felt abused, strung out, and on the verge of collapse. The steep slope would need grading, but that was not allowed because of permitting requirements that restrict the stirring up of contaminated land. A foot of standing water with a seeping oil slick on its surface covered the only usable flat area at the site.

Although Seann and I pride ourselves on taking on both troubled land and troubled individuals, we recognize that there are some plots of land, and some people, whose challenges and barriers are too daunting to work with. This land was beyond our ability and resources for recovery, not too dissimilar to the worst crackheads or junkies we encountered daily floating down the streets and back alleys only blocks away.

Telling city officials we could not use the land was difficult; they had put in a lot of work to make it available, and we came off as sounding unappreciative. In the end, however, it was clear to everyone that the site needed a kind of remediation we could not offer.

Our search for land in the city led me to other such emblematic and symbolic observations, reminding me once again of the abuse that so much land has been subjected to. Almost every piece of land in every city across the world has been exposed to some form of contamination. As a society we seem to know about maintaining homes, buildings, and roads, but we have little knowledge of how to care for and nurture land, whether farmland or forest, prairie or desert or wetland. We see land only in the value it renders when cleared or sold or cut or paved or built on. We have forgotten that each of us is inextricably connected and dependent on living soil for our nutrition and our health, and so we treat soil like dirt.

A friend of mine says that by paving over our precious soils we are preserving them for future generations. This is little consolation when you consider that we have paved over and built on some of the richest and deepest topsoils in North America.

I never thought I would find myself seeking out paved land as the most desirable place to run a farm, but in the end I concluded that pavement provides the best separation from what lies beneath, which in a city is almost always soil that is unsafe for growing food. Pavement also provides an easy surface on which to implement our raised-box movable farming system, which isolates the growing medium from native soil and has been the cornerstone of our model.

This raised-bed system, along with a lease document we developed that appealed to both private and public landowners, opened doors for us. The language in our lease demonstrated the portable, short-term, and seasonal nature of our operations, provided information on the property tax relief benefits for leasing to us (depending on the classification, up to 90 percent potential reduction), explained our ability to vacate on short notice, and promised that we would be fully insured with public liability and tenant property coverage. It also addressed the aesthetic improvements and positive corporate image that would come

Urban Challenges and Urban Advantages

Farming in the city presents unique challenges that provide the basic context for this work. These challenges are common to every city everywhere, and they inform the work we do agriculturally. Here are what I consider the most common urban farming challenges:

Challenges

- Limited lateral space
- High land values
- Contaminated soils
- Theft and vandalism
- Pavement
- Loss and damage of crops from birds and rodents
- High costs (water, infrastructure, permits, housing, et cetera)
- Lack of experienced skilled labor and management

And yet I'm also convinced farming in the city offers distinct advantages for the farmer and for those eating his or her food that far outweigh the challenges.

Advantages for the Farmer

- Proximity to markets
- Proximity to a large customer base and labor
- Lower weed (and some pest) pressure
- Warmer conditions, due to the urban heat sink, which allows for earlier production

Advantages for the Neighborhood

- Direct visual and participatory connection to the farm and the farmer
- Improved food quality
- Access for children to food and how it is grown
- Greening of neighborhoods
- Job opportunities

from association with us, along with the inherent goodwill their lease to us would generate in the community.

Developing the lease was one thing. But as Kira and Seann discovered, the permit process for starting a farm in the city is cumbersome, incredibly time consuming, and expensive. The regulations and codes that guide the process can make it impossible for a small farming business in the city to ever get started. The one-size-fits-all reality of municipal bureaucracy may work for someone building a home or an apartment building or a retail store, but the profit margins for small- and medium-scaled farms will never be on parity with other similarly scaled businesses, and as such the time and expense of obtaining permits can be prohibitive.

Accessing our original farm site on the parking lot next to the Astoria was much easier than anything that came after. That parking lot is leased from the Sahota family, their agreement heavily influenced by the fact that they realize a significant annual savings in property tax benefits from having us farm that site. This savings is based on a reclassification of their parking lot from commercial "parking lot" to "park." That original farm site—our spiritual home—firmly placed Sole Food into the heart of the neighborhood we were serving and became the foundation for our project in its beginnings. And while we have never been able to get the Sahotas to sign an official lease, we have been able to continue farming that site relatively unfettered.

Land in this city is extraordinarily valuable, and nearly all of it's been gobbled up. Most of the available land on offer by the city of Vancouver was either covered with rubble, highly contaminated, or too small for commercial consideration. We needed sites that were at least one acre in scale. During 2011 we got to know every former gas station, abandoned and demolished building site, open parking lot, and rooftop that existed in the city. We eventually settled on a few sites, but not without a few more false and sometimes painful starts.

By the spring of 2012, after a mountain of documentation, phone calls, emails, city council meetings, and permit hurdles we had finally secured a one-acre site in an industrial zone, another acre next to a homeless shelter on a busy corner, and a small piece of land formerly the site of the 2010 Winter Olympic village, all owned by the city of Vancouver. We were also able to secure a lease on two acres of a five-acre parking lot privately owned by Concord Pacific, a high-rise developer with projects across the country.

———

Our one-acre site at First and Clark sits below an overpass in an industrial part of town across from a fish processing plant, flanked on one side by the Vancouver school district maintenance yard, and on the other by an entire ecosystem created by the bridge that towers overhead. There are depressions in the giant concrete support columns that hold up this bridge, and people had established rustic shacks there, hidden from view. But the understory of a highway overpass does not just attract homeless people seeking a dry place to sleep; there were huge truck and tractor tires and an array of industrial artifacts.

We filled a Dumpster with the pack-rat accumulations left by the temporary residents who had passed through the farm site.

Despite the detritus, when I first saw this location I knew it would work well for us. It is perfectly flat, already fenced, two-thirds paved with a slope behind it that I felt could become an orchard or serve as a buffer from traffic and the urban world.

I generally welcome sites that are in high-profile neighborhoods with good visibility. Our goal was never to do this work quietly or privately; we wanted the world to see the work we were doing, to see what was possible. And yet some farm installations do not lend themselves to public view, such as the high tunnel greenhouses that we need for season extension, growing heat-loving crops, and growing the transplants to support our broader farming operations. The gothic-shaped steel frames covered with plastic have always appealed to my aesthetic, but most urban dwellers would not want them in their viewscape. The First and Clark site is not very visible, and its industrial neighbors don't expect a pretty view. As it turned out, satisfying the visual sensitivities of neighbors would be the least of our challenges.

The fifty-five-gallon drums bearing chemical names so complex I could not pronounce them should have been fair warning that despite its obvious advantages, this location, once home to a gas station, had some significant hidden problems. Installing sixteen thousand square feet of greenhouses in the city, when no such building codes existed, and on land that we discovered was highly contaminated, proved to be challenging.

As part of our permit process, the city required that we produce an environmental plan that would outline how we were going to deal with contamination and runoff—problems we did not create but that existed from the tenure of the previous tenant, a well-known oil and gas company.

Legal and liability concerns on the part of city officials were the motivation for their request for the environmental plan. The city didn't want their former oil-company tenant to find a way to get off the hook for contamination it had left behind. The city was concerned that if Sole Food moved in and started farming without a baseline environmental study, there was the real potential for the prior tenant to claim it was no longer responsible for contamination it had created.

I'm not sure how our society has come to think that extracting, destroying, and polluting for financial gain is acceptable, and you will not be held responsible or accountable for the results. My parents might dispute how successful they were in requiring me to do the dishes after a meal or clean up my room on a weekly basis, but the basic lesson was always there: "Michael, clean up after yourself!"

Sole Food didn't have the kind of financial resources required to produce an environmental plan on the level the city required, but a local firm received a brownfield remediation grant from the Ministry of the Environment to produce an extensive study and set of recommendations.

We submitted the fifty-eight-page environmental plan along with an operations plan outlining what we would be doing on that site and how, a set of architectural drawings of the tunnel houses with engineering specifications, and our insurance documentation for the site.

Fulfilling endless municipal requirements absorbed staff resources and money and delayed the real work of planting crops. It was useless to try to explain to city officials the realities of farming, the urgency of spring, that we had staff hired and plants waiting.

In a climate that only provides a seven-month window to crank out long-term crops such as tomatoes and peppers, losing a month or two can be disastrous. That first spring on this new site was vanishing, and we desperately needed to plant the now leggy, past-due tomatoes, peppers, and eggplants that were becoming as stressed as we were, waiting for the wheels of the city bureaucracy to grant us permission to begin.

We were about to lose ten thousand pepper, eggplant, and tomato transplants that were taking up valuable growing space on my family farm on Salt Spring Island. Members of our staff were already on edge. Everyone wondered whether starting more farm sites was even a good idea; whether they would still have a job if the building permits did not come through or the plants did not survive. Regardless, we knew the tunnel houses needed to be installed, so after being advised off the record to "do what you have to do" by a city official whose name I will not reveal, we hacksawed the lock off the gate and started assembling and installing four 20-by-200-foot tunnel houses on top of that former gas station.

Like most of our efforts to create farms on pavement and within the city, the unexpected challenges and delays were numerous. The ground that was not paved was uneven so we leveled it with sand, and much of the site was asphalt or concrete so we had to jackhammer holes to put the posts in. We were not supposed to be compromising the ground by disturbing it in any way, but there was no way to level and anchor the tunnel houses without doing so.

It required a crew of ten people and three weeks to get sixteen thousand square feet of steel framework up and secure, but covering the houses with plastic and completing the job would have to wait.

Someone must have complained or notified the city of our work on the site, so they came to inspect, saw the framework of the tunnel houses now in place without a building permit, and issued a stop-work order. It would only be lifted several weeks later, once the permits were issued.

When we eventually got back to work on that site we discovered that $5,000 worth of greenhouse parts that we needed to finish construction had been stolen, and someone had taken up permanent residence in the back of one of the tunnels. Replacing the parts and evicting the squatter delayed our work even further, but eventually we completed the job.

We've all heard the cliché that it is easier to ask for forgiveness than it is for permission, but in this case we needed both. And yet I know for certain that had we asked for permission rather than forging ahead and installing those tunnels, to this day we might not have a single plant growing on that site.

We hauled those plants to the city from my farm on the island in an enclosed bobtail truck. As we loaded the flats onto the truck I wasn't sure what the outcome would be. Would this become a funeral procession or a revival parade? We had definitely pushed the plants' survival to the edge. Unloading that truck in the city, I could see the expression on the faces of our crew quickly shift from excitement to concern.

In the end it was a profound and positive visual lesson for everyone, seeing those spindly, stressed pepper and tomato plants, leaves sparse and yellowing, plants that had been held back and abused, become healthy and highly productive when nurtured and given the right conditions. I felt so much was riding on the survival and recovery of those plants, not so much because we needed the food and the income they would produce, but because I saw those plants—like the people we were working with—stressed and needing nourishment, nurturing, and love. When the plants recovered—and they did with amazing speed and gusto—I felt a great sense of hope and possibility.

Before long, the city of Vancouver also got over its stress, and we all shared the credit for a national award we received for that site. The Brownie Award, created by the Canadian Urban Institute for excellence in remediation of brownfield sites, was presented to the consulting firm that wrote the plan, for our innovative work in making safe use of that contaminated site.

———

Concord Pacific, one of Canada's largest developers of high-rise office and residential buildings, is a very different kind of private landlord from the Sahotas, our landlords at the Astoria. Walk into Concord Pacific's "presentation center," in a modern glass-and-steel single-story building overlooking False Creek, and you are greeted by two elegant women behind a sleek reception desk. This is where prospective buyers come to view models of upcoming developments and existing units for sale or rent. The feeling is of luxury and style, a well-crafted environment radiating comfort and modern convenience.

Concord Pacific's landholdings in the central core of Vancouver are substantial, and its founder and president Terry Hui is a brilliant Stanford engineering graduate who can appear more like an absentminded professor than the successful developer he has become. Our first benefactor, Frank Giustra, introduced us to Terry.

Both men are serious power players in their own right, which, from my perspective, represents a reality that cuts both ways. They both do a lot for the benefit of the community in Vancouver and beyond, but I cannot ignore my own internal conflict at how much of my energy and effort I devote to reaching out to the wealthiest citizens of this city to ask for a little help for those who are the poorest. It's not that I am not thankful. I am, and I know that I continue to need their help. Over the years I've come to understand that the powerful and financial elites of the world, especially when driven by cooperation rather than competition, have an astonishing ability to make things happen. In a single phone call they can instantly produce results that others spend years trying to accomplish. Both Frank Giustra's introduction and Terry Hui's near-immediate understanding of our goals gave us access to what became our largest farming site in the city and the new center of our operations.

That first meeting with Terry in February 2012 required no more than half an hour. As soon as the meeting was over, accompanied by representatives of Concord Pacific, we walked a short distance to see the parking lot at the corner of Pacific and Carrall that would later that spring become our farm. This was the easiest audition I have ever had. We started with two acres, agreed on the princely sum of $1 per year, discussed some basic principles for the lease, and that was it.

As Seann and I were leaving, we shared a glance that communicated our disbelief that anything could be that simple, but in the end, compared with the mountain of challenges we dealt with on a daily basis, it really was.

Our relationship with Concord Pacific has continued to grow. They have been incredibly generous and supportive in many ways, and we endeavor to be good tenants and neighbors. This type of collaboration between unlikely partners is what the world needs more of, and I realize how much our common future depends on it—the ability to bring together disparate forces to cooperate.

And while our Sole Food farmers do the physical work of growing food and providing employment, we see urban agriculture, when done well, as the ultimate social, political, nutritional, and financial collaboration between municipal government, private and public landowners, foundations and individual funders, and the community of eaters we supply.

We cannot do this work alone, and it may be that this kind of farming is different, less because of its urban context of parking lots and chain-link fences, and more because that context requires a whole community to make it work.

In hindsight, the major adjustment to our expansion plan, reducing those original thirteen sites to five, turned out to be a blessing. Aside from financial considerations, we would have been stretched too thin and would not have had the management capacity to hold thirteen sites together.

As it was, the expansion from one site to three in spring 2012 put a real strain on our staff, to say nothing of the strain it put on me and Seann. Attention that had been going to supporting our people had to shift to managing massive installations, neighborhood and city politics, and planting and harvesting from nearly four acres of pavement. We realized too late the toll this took on some members of our team whose capacity for change and transitions was already limited.

———————

Diagnosed as bipolar and with borderline personality disorder, Rob was living at Union Gospel Mission and in treatment for drug and alcohol addiction when he attended a job fair at the Carnegie Community Centre. We hired him that day, and he helped build the boxes at the Astoria farm.

The first time I met Rob, I could feel him checking me out, trying to get a handle on whether I was for real or just another do-gooder passing through. Rob is a big man with a round face and bright-red complexion. He adorns himself with beaded necklaces and often wears a bandanna or scarf around his head to hold back his shoulder-length blond hair. On first meeting him, his size and his who-the-hell-are-you expression were unnerving and made me uneasy, but after an introduction and the start of a conversation, it became clear that this guy had incredible intelligence and could be warm and engaging.

But Rob could also completely withdraw. I remember showing up one morning at the Astoria location to talk with him. He was lurking behind a greenhouse, and his energy was so dark and thick you could feel it from across the farm. I made a feeble attempt to approach him, but it was like cornering a wild animal, and I got the feeling that he would snap and attack if I got too near.

Rob would eventually come around, lift himself up, and be right there and fully engaged in the most intricate details of the work we were doing. He took to the farm and the concepts and philosophy of good farming with gusto and immediately started consuming every piece of literature and information on food and farming he could get his hands on. I've never studied farming in the formal sense; I've come to it through practice and experience. But Rob wanted to know in-depth technical details that I could not always provide. I would have a discussion with him on the farm about the role of calcium in the soil or optimal soil temperatures for seeding different crops, and the next day he'd have done his research and come up with a new set of questions and ideas. His mind was always fully engaged. I love the banter that can take place when someone digs into the incredible complexity and depth of the work that we do with soil and plants and food, especially when he is new to it and his desire to learn is so acute.

Early on Seann had the idea that Rob might have what it takes to do some management and direct other people. Seann was still employed and working part-time at United We Can's bottle depot, and we needed someone to fulfill a day-to-day managerial role. When we expanded to multiple farm locations, Seann put Rob in charge of two farms. My gut told me this was not a good idea. I understood Seann's instinct—give someone responsibilities to grow into and you support his growth. But giving Rob all that responsibility never felt entirely comfortable to me, and I disagreed with the decision.

As happens in any partnership or collaboration, Seann and I have not always agreed. In the beginning, especially, there was a lot of give-and-take and letting go. Although I cannot imagine a more compassionate person to collaborate with, there were occasions when I had to accept Seann's youthful arrogance and his desire to blast forward too quickly to make things happen. In my thirties and forties I was no different, but my own mistakes had shown me the consequences of moving too fast.

It was good for me to learn to let go of having to be right and, as others had done for me when I was younger, allow things to find their own conclusion. Unfortunately, the results of some hasty decisions backfired quickly, dramatically, and often expensively. But there were no I-told-you-sos, nor were there immediate acknowledgments of mistakes made. We all just moved on.

Rob could be so incredibly caring for the rest of the staff, but he could also be abusive and, for lack of a better word, a little *off*. Seann and I have talked in hindsight about that transitional period and what went wrong. "At first it didn't seem like there were problems," Seann has explained, "but at the end of the first season he kind of fell apart. It's kind of shocking to see somebody unravel like that, because he doesn't wear his addictions like other people and because he is so smart. You don't think somebody who is that smart can fall to such depths. It's a prejudice, but we all have it. I think that's what made it harder to deal with Rob because you think, *You're too smart to be doing this*, but he has no control. When his medications were changing, or who knows what, he would go off track."

Eventually on a Saturday morning, in the summer of 2012, at 4:45 AM, Seann woke up to go to the farmers market and found a series of text messages with photographs on his phone. They were of Rob cutting his legs and bleeding into a bathtub.

Rob survived this incident, but it was the end of his work with us. Soon after, Seann and I went to visit him at his rented apartment. The shades were drawn, the room was dark, and Rob was sitting alone drinking cheap wine and taking OxyContin. We brought lunch with us and sat together eating and sharing stories.

Rob

Rob is probably the most well-read individual I have ever met. I'd come to work at the Astoria farm and he'd be rattling off a litany of literature he had consumed just in the last thirty-six hours. At first it was hard to believe that someone could read and retain so much. And so while we were sitting with him in his apartment I challenged him, tried to come up with a random obscure title that he had not read. I started with a travel theme—one of my favorite books, Bruce Chatwin's *Song-lines*. Not only had he read it, but he pulled a copy from his bookshelf. Peter Matthiessen's *Snow Leopard* was next, Bill Bryson's *In a Sunburned Country,* and on and on. He'd read them all. At the end of this little game Rob handed me a consolation gift, a wonderful book about Mexico.

It has been a few years since Rob has worked with us. Seann and I have had lunch with him on a couple of occasions. The last time I saw

Rob we met him at a local coffee shop. He was wearing a pair of shorts and had a boyish haircut, and if it weren't for his sheer size and the scars on his arms, you might think he was twenty years younger than he is.

"It was the best work experience I had in my whole life," he told me during that meeting. "I never woke up thinking, *I don't want to go down there*. Then my mom died, and I made that same choice I always do. I don't have a whole lot of resilience. It's like a skip in a record that always goes back to the same song: booze and pills. Sometimes the bravest thing you can do is to get out of bed."

Rob helped shape that farm. And by his own account, Sole Food did its work on him too: "I am proud to say I was part of something from its genesis to its teens. There was this one morning during the second season that says it all. It was 4 AM and I was standing on the farm in the middle of the city and I looked around and I felt like I was communing in church."

Stories like this this one from Rob are bittersweet for me. We build something like church, and offer opportunities for community, but we cannot save people. I have to remind myself that neither Seann nor I is a social worker or therapist; nor are we physicians or psychologists. Indeed, I am often asked if we have a social worker on our staff or which one of us is trained to help the range of folks who seek work at Sole Food. And

while we would love to have the financial resources to hire such a person, we see therapy happening in other ways: harvesting carrots, seeding salad or radishes, preparing the soil for a planting, hearing a loyal customer at the farmers market express pleasure in the foods we provide.

We believe in the power of growing food and nourishing others as a way of healing ourselves and our world. Our job has been to set the table, provide the foundation for something to happen, based on the belief that the simple act of planting a seed can bring new life to the world.

Salvation is a big idea—no one does it to you or for you. I never got into this work to save anyone, but I believe that simple things can help in profound ways. I know from experience, for instance, that salvation can come from the soil. It happens when you get your hands in it, when you learn how to maintain its nutrition and biology, when you cultivate and keep it open and oxygenated. It's no different for you or me; to stay well you need to be well nourished and hydrated, to breathe deeply, and to stay open. These are basic principles for human health and for plant health as well.

And so I farm because I want to eat well. I also farm so I'll have stories to tell. What's more, farming, like storytelling, keeps me sane. I've come to know that if you provide someone with a living organism to care for, something that she cannot leave alone even for a day, something that depends on her for its survival and well-being, then you've given that person a reason to live and stay well. She will have to be present, pay attention, and step away from her troubles, because there are plants and animals and people who depend on her.

Sole Food was built on this very simple idea: rehabilitation through meaningful work. Growing food is high on the list of meaningful work. It brings us down to real basics; it allows us to see cause and effect play out daily, to see how our action or inaction creates interaction. We feed the soil that feeds the plants that feed us. We begin to sense how working with plants and soil and sunlight and rain and wind can heal us.

Good farmers know that to be successful we must be immersed in a consummate love affair—with the land, with the living soil, and with the broader community that eats our food. We know that this work alone will not heal the folks who come to work with us. We have no illusions that a farm or good food or real work will help everyone get healthy. But the raw biology, the simplicity and physicality of the work, the magic of seeds emerging, plants thriving, food being harvested and filling boxes and bellies, in places where previously there was nothing but hardscape and trash and rubble: This is recovery. Experiencing it will touch and change you.

CHAPTER 4

Walking the Land

Providing context for the work we do, teaching the hows and the whys to our staff, sharing technical and philosophical details on every step of the process has been central to Sole Food's mission. And so, each year my friend and colleague Josh Volk travels from his home and farm in Portland, Oregon, to Foxglove Farm to join me in teaching our annual Growing for Market workshop. We've known each other for twenty-five years, having met when he and his partner, Tanya, attended one of the Food for Thought discussion groups held in the late 1980s at the farm I was running in California. Josh brings his background as a mechanical engineer to farming. Although he is seventeen years younger than I am, I've learned a lot from him about a systems approach to farming. Josh does farm planning on Excel spreadsheets; I use narratives and lists. He plots and plans; I operate more intuitively. The blending of our farming and teaching styles provides a nice balance for workshop attendees.

Since the inception of Sole Food Street Farms, Seann and I have gathered our staff together for the annual trek to Salt Spring Island for the Growing for Market workshop. We travel by truck or car from the Downtown Eastside to Tsawwassen to catch the ferry to Swartz Bay on Vancouver Island. We hop a second ferry to Fulford Harbor on Salt Spring and once there take a half-hour drive up and along the road that winds up Mount Maxwell to my farm.

Though circuitous, the path to the island is quite well trodden. Every summer hundreds of urban refugees escape the stress and clamor of their city lives to come to our farm refuge. They stay in cottages, cabins, and the original log house, receive baskets of fresh produce, eggs, and meat delivered to their doors from surrounding fields and pastures, and cook and hike and swim. Many come to the farm to participate in one of our workshops—farming, cheese making, canning, forestry, mushrooms, beekeeping, and more—or to take part in our annual festival or our Farm, Arts, and Culinary Camp for kids.

Organizing twenty people from the Downtown Eastside for this journey, however, comes with complications. Sleeping bags have to be obtained, special arrangements secured for methadone or morphine prescriptions normally provided just a day at a time. We have to round up our people and transport them.

And what's more, unlike those who flee the city once a year for the calm of Foxglove Farm, for many of the Sole Food crew leaving the city isn't always a stress relief. Despite the harsh realities that exist in their neighborhood, it is their home and provides a support network that can be difficult to leave behind. My island farm is dead quiet, pitch black at night, with no traffic, few people, and no 24/7 easy access to anything from anywhere. For some of our farmers this can be terrifying. There is always someone who wants to go home as soon as he arrives.

Still, it's amazing how quickly the layers of tension and fear can drop away and this foreign rural world can become comfortable. By the end of the three-day workshop there is a noticeable settling and sense of calm among our staff. By the time they are leaving their faces have relaxed and they look different. I have often felt that we should photograph folks when they arrive and when they leave—*Before. After.*

The farm is the 120-acre remnant of one of the oldest homesteads on the island. You can feel history here under your feet, you can see it in the old buildings and taste it when you eat an apple from a tree planted by John Rogers, an immigrant from England who first settled on this land at the turn of the last century. We think of those early pioneers as hardscrabble, just trying to survive, but this land was cleared and developed with a well-planned aesthetic, the fields surrounded and intersected by swaths of intact forest. John Rogers and his wife, Alice, arrived on horseback in 1897 and began gradually clearing the land to build the original log barn and house. They planted an orchard, potatoes, peas, and cabbage and raised some pigs and cows for meat and for milk. As their family grew, they added other dwellings until the whole scene evolved into a little village beautifully situated in the heart of the valley.

From March through October those fields that Rogers cleared are still incredibly active, bursting forth with sixty different fruits and vegetables: melons, asparagus, strawberries and raspberries, blueberries, carrots, beets, greens, beans, and corn. Larger fields of grain, hay, and pasture support a flock of laying hens and a few pigs. The intensive cropping systems, the animals, the equipment and tools, and the greenhouses all provide a visible working model that give the Sole Food team and other workshop participants context for the farming concepts Josh and I present.

It is a beautiful September evening as Josh and I assemble the group in a restored building that was originally a chicken coop, before being converted to this classroom space. We begin the first evening with introductions, each person explaining a little about where they come from and why they are here. Our Sole Food participants shuffle into this space hesitantly, a bit awkward, a few of them late, and sit at the back of the room. Their introductions are humble, a glimmer of pride revealed only when they describe themselves as a Sole Food "farmer."

Kenny introduces himself, at first a little shyly, but then he opens up and talks about how at first Sole Food was just a job, but after only a few years it has given him a new identity and real purpose.

Donna's introduction is brief and to the point. "I'm a farmer now, and I'm here to learn more about farming."

There can be sixty or seventy participants in these workshops: wannabe farmers, people who are already farming but want to refine their skills or get some new inspiration, landowners exploring possibilities,

Foxglove Farm

and urban dwellers who are seeking an alternative to their office-bound lives. Everyone else who attends comes from very different backgrounds than our co-workers from Sole Food.

After the introductions I launch into my ideas on how to begin creating a vision for a farm. I reflect on the starting points for good work in the world, what it is that stimulates us to initiate something, where ideas emerge from, and how we move from idea to real-world application.

This first night I offer an overview of what's going to happen in the next few days. But I'm also trying to encourage everyone to think about the next few months and years and beyond.

"We'll do a guided farm tour first thing in the morning," I say, "followed by the business planning section. Then we'll move into the foundational session on soil fertility. Josh will do a crop planning session, and in the afternoon we'll get into propagation and field preparation, and we'll be out in the fields looking at planting, seeding, cultivating, and irrigation."

I can almost hear a sigh of relief from the group, hearing that we will balance all the talk and sitting with hands-on field explorations.

"We're going to show you everything from how to stand when you're working to what tools are appropriate to use when. On Thursday we'll get into marketing. We're going to ask for three volunteers to present their farm ideas, and we're going to analyze and break down each one for you. This will be followed by a short presentation on pests and diseases, then season extension, harvest and post-harvest, and dress for success. Thursday afternoon we have a section on crops A–Z that gives you an opportunity to ask questions about specific crops.

"But again, what's really important here, or in any endeavor, is to have a vision to start from and an understanding of why you are here."

"I always begin," I continue, "with a series of questions. Why am I doing this? Who am I doing it for? What are my goals socially, personally, financially? What skills will this require? Who will it benefit? How will it be funded initially, and what is its longer-term potential?"

The questions are where I always begin—in my projects as well as in my presentations. Specific questions help me to consider my intentions, understand my motivations, and articulate my expectations. The impulse is the same one Seann and I acted on while brainstorming during the early days of the Astoria farm, huddled in the Wilder Snail coffee shop, away from the sogginess, seeking clarity. Questions help me to consider what is expected by and from the broader community; they help me see what's at stake even with our own families.

"These questions," I say, "often have no immediate answers. The answers will reveal themselves in time and as the project develops."

My early planning often leads me to indulge somewhat in esoteric realms; it is the time when I give my mind permission to take off into unlimited creative possibility. While I have been known to consult an astrologer or refer to the phases of the moon for optimal planting times, I offer a disclaimer that reinforces my pragmatic nature—I do not consult the oracle, have my tarot cards read, or rely on psychic council for myself. Only for the crops and the fields.

"My approach to beginning is quite simple. It involves a piece of paper, a pen, a quiet space, and some time without distraction."

There is some subdued chuckling in the room when I describe these ancient non-digital tools, as if I've suggested that they use stone tablets and chisels.

In a vulnerable moment, I describe my discovery, as a late teenager, that writing down my dreams and desires could open doors; opportunities would emerge. When I was younger those dreams were simple—a car, a job, or a girlfriend, someplace I wanted to travel to.

"My desires evolved as I got older—I wanted a place to settle, a family, farms, food, public education, social service, and a community to provide for and to work with. I've written all this down in my life."

I stop and look around the room as I describe all this and I consider how abstract these ideas might be for Kenny, Rob, Donna, and other members of our crew from Vancouver. Is it a luxury, I wonder, to be able to look beyond the present moment, beyond food or housing or relief from gnawing habits? Not everyone has the freedom to make these kinds of choices. To consider, plan, or design a future may be another form of privilege, I think, as I notice some of our crew fidgeting in their seats at the back of the room.

Speaking up again, I reflect on the qualities of the intimate relationships most of us have over the course of our lives. For my part, I've been blessed with two very long-term relationships, one of which continues to thrive after twenty-seven years. I too have participated in serial relationships, but mine have been with land and with farms. "Remarkably," I tell the group, "each farm has had an eerie resemblance to the scribbled piece of paper I had written some time before and stashed away in a drawer."

I know it has become my refrain, but my visions and successes have always come down to the details, I remind the room. All great ideas depend on the little stuff. When the only thing you have at the beginning is an idea, best to describe it in as much detail as possible, I suggest.

Here, I say, is what an early agricultural plan can look like for me. I use my mind's eye to imagine what the farm will look like in three, five, and ten years. I write this down. I describe the neighborhood or community, the water sources, the soils, the scale, infrastructure or buildings. More notes. I consider whom I want to be working with, how those relationships will be structured—as a partnership, a co-op, a benevolent dictatorship. I then describe how I would like to distribute or sell the food. When I am really present and clear, my descriptions are highly visual and detailed.

The result is a simple document that becomes my map, my itinerary, my ticket and passport for the journey that is about to begin. Some of the group listening have begun taking notes.

When Seann and I created this document for Sole Food, we did not have the land, the money, or the farming sites, but all kinds of doors opened up after we articulated our vision. Together we built that entire farming enterprise on paper, and when we were done we put the plan away. Our first major funder soon appeared. I remember on many occasions the look of doubt I saw on Seann's face, and while he may not admit it now, I am sure he wondered how all the plans we made together would ever come to be.

I have often thought back on those innocent days sitting together, scheming and writing down what appeared at the time to be a wild and crazy and somewhat impossible dream.

"Nothing put forward in this visioning stage is set in stone," I remind everyone. "Everything can change. In fact, everything will change."

"I know some people have been afraid to do this exercise," Josh interjects, "because they feel like once they have written their idea down, it can't change. But as Michael says, it's not set in stone."

This is the first seed that I plant, the one that nourishes the whole endeavor. It is how I initiate every project. The bridge between visualization and manifestation is, simply, hard work. These things do not happen on their own.

As we are closing the evening session Josh and I give an assignment. We ask the group to go back to their cabins or tents and consider their own vision. We ask them to write it down and bring it with them the next morning.

"When we walk around tomorrow on our tour," I say, "I am going to be describing some context, some history, and what's happening in these fields and orchards today; I am also going to try to describe what was going on in my head when I started and what I'm seeing now. And as we are walking together I want you to put yourself in my head or the heads of any of my crew and imagine you're the one making decisions. What would you do?"

We disperse, flashlights like fireflies bobbing and moving in different directions across the land. I find my way back up to my home in the darkness, feeling my way along familiar routes, listening to distant conversations. I consider this group, the range of backgrounds and experiences, and I wonder how I can possibly teach to such diversity. And then I remember that we are on this amazing land, with good farming all around us, and I am comforted by knowing that in the end it does not matter so much what I say as what they see.

The next morning we gather for a self-serve breakfast of granola, muffins, yogurt, fresh strawberries and blueberries, and tea and coffee. We move outside for a check-in, an introduction to the day and to the farm tour I will be leading.

Most folks are bright-eyed, ready and eager; a few are still groggy. It's an unexpectedly chilly morning, and everyone is shoulder-to-shoulder, wearing every layer of clothing they could find. I remind the group where the outhouses are, a few rules about smoking and dogs and smoking dogs, about mealtimes and where to park. I find myself taking the attendance of Sole Food staff in my head. I quickly scan the faces of those who are present. They stand in the back of the group, slightly unsure, but a little more relaxed than on their arrival last night.

I begin the tour with a little history, about the native people who moved through this land fishing from its creeks and lake, how the original homesteaders encountered land completely forested with trees too large in diameter for their crosscut saws, how they augered holes into the sides of the giant trees, filled them with coals, and burned them to the ground. There is some shock from a few people in the group when I describe this practice, but it was the Rogerses' hard work in clearing this land that created the open fields and the foundation for the work we now do.

Maybe it's my projection, but I suspect the people here from Sole Food know all too well how destruction and recovery are part of the same cycle. The realities of farming offer so many metaphors like this—life and death, decay and fertility, destruction and recovery—for those we work with in Vancouver. I go on to describe the bootlegging operation that existed here, how there were extensive orchards and fields of rhubarb exclusively devoted to hard cider, and I am reminded that every generation has its own particular forms of intoxication; its forbidden substances; its bootleggers, its dealers, and those who succumb to and are ravaged by overindulgence and addiction.

We walk from field to field, the group fanning out a little behind me, and I talk about the philosophy behind our practices, how we designed and built the farm, why we chose certain crops and details on how we grow them.

I reflect on our thinking when we took over this farm, that we waited a year before doing anything to just watch how the sun moved across the land, which fields dried earliest in the spring, where the cold pockets were and what areas had the best exposure or stayed warm. That first observational year allowed us to see what weeds predominated where and why, the differences in soil fertility from field to field, the impact of a long history of grazing animals on the farm, and why the original homesteaders chose certain places to plant their orchards or site a building or simply keep some land untouched and pristine.

I describe the crop mix we decided on, how those decisions were influenced both by what we like to eat and what was not already being grown in any quantity on the island. In those early years my vision was to make this a mid-scale farm—to step away from the smaller boutique production that most of us had been doing for the farmers markets and restaurants and grow for the supermarkets. But in the end the dearth of skilled labor kept our production levels in check.

As we make our way through the vegetable fields, I pick and open a few swollen bean pods, revealing tiny lime-green kidney-shaped flageolet beans, pointing out how they are only days away from harvest. I discuss strawberry varieties, how we have been experimenting for years to find the perfect ever-bearing variety. We sample a few, and I discuss the subtle differences in color that determine a truly ripe one from one that will be mediocre.

"Check it out," I say, having picked three berries. "Which one do you think is ripe?" One of the fruits is red, the other has a white tip, and the third one has gone beyond red to crimson. The differences are subtle, but it is the crimson-red fruit that will have the most flavor and sugars and perfect texture.

Harvesting, I remind the group, requires a good eye, a million subconscious decisions made every second as you navigate your way through a field deciding who is ready and who must wait. We harvest a Golden, a Chioggia, and a Merlin beet, and as I cut into each I reveal and describe the visual benefits of beet bouquets in the market, multiple colors in one bunch versus single-variety bunches.

Arriving at one of our tunnel houses devoted to melons, I launch into descriptions of one of my favorite crops. "Melons have no humility or self-control," I tell the group. "They know no boundaries and will take

over whatever is around them." The real estate in the steel-framed tunnel houses that provide protection and a little extra heat is valuable, so we train our melons vertically to use the space more efficiently. Everyone looks on as the fruit hangs on vertical vines supported by hairnets so they don't break the vines or pull them off the strings.

This stop on the tour stimulates a long discussion on the values of tunnel house infrastructure and season extension. "There is probably no other single piece of farm infrastructure that is more important than the tunnel house," I remind everyone. "They provide essential season extension, allow us to grow crops that might be marginal in this climate, provide protection from insects and disease, and keep the rain off moisture- and disease-sensitive crops."

We eventually make our way back to the classroom space, take a few moments for tea or coffee, and settle in for a day of instruction and demonstration.

I start off by reminding everyone that while what we just experienced is a large open-field rural farm, many of the principles and practices they've seen are applicable to the city. And yet we must bear in mind that carrying out these principles in the city requires careful consideration of the differences between open-field agriculture and what can be done in a parking lot, between buildings, on top of contaminated land, in a populated neighborhood, behind fences, and without the space and the broader biology of an open field.

When most people close their eyes and think *farm*, they see something very much like what surrounds us here today: a pastoral scene with animals grazing, a farmhouse, wood fences, fields of crops, a woodlot, a meandering creek, maybe an orchard. This is the classic, culturally embedded image we all have of a farm in rural landscape.

A farm in the city asks us to imagine a very different scene. And while many of the practices may be the same on an urban farm—preparing soil, cultivating, irrigating, harvesting—the strategies required must accommodate a different set of challenges and circumstances, such as high-valued land, depleted and poisoned soil, paving, and extreme space restrictions.

I had a seven-year courtship with this land on Salt Spring Island; it took seven years of visiting and contemplating a purchase before we bought the property, tied the knot, and moved in and started farming it. There was a powerful attraction and chemistry, but I wanted to know that our connection was deep and that it would last. I approached this land like I do most land, with an eye for existing soil fertility, drainage, weed pressure, exposure, wind, access to water, and proximity to supplies and to markets.

Farmers train themselves to read land, understand what it means when certain weed communities dominate, see how well those weeds are growing, and assess the smell and depth and feel of the soil. When we started on Foxglove Farm at twelve hundred feet above sea level, on the side of a mountain, there were pockets of fertility from years of permanent pasture and animals grazing, but the soils in many fields were rocky and thin, the existing weed growth weak. Now there are fields where you can reach your arm in almost up to your elbow.

Knowing how to read land is valuable, but when choosing a location and considering a farm in the city, other instincts must kick in. And first, those instincts must be developed. In the city we seek out fences for security from theft and vandalism; our land cannot be shaded by high-rise office or apartment buildings and must be easily serviced by trucks or forklifts. It must be relatively flat, and since we assume that the native soil is not usable, we prefer land that is "capped," or paved. We knew some of these things when launching Sole Food. But, I admit to the group, some of our education came the hard way.

Here is a basic principle I've known as long as I've been farming: Wood and water don't mix. Water always wins. In Vancouver we have lots of it. It falls from the sky for eight months of the year, not in a deluge, but in a constant, ever-dampening, never-ending sop—what the more optimistic people in this part of the world call liquid sunshine. And while our part of British Columbia is a world of water, it is also a world of wood, with boreal softwood forests thriving in the wet conditions. When we think of constructing anything in these parts the go-to is always wood.

We built our first farm site at the Astoria with Douglas fir two-by-four framing and two-by-six plank walls. That first farm was small, so we could afford to use wood thick enough that the eventual rot and decay would take years to manifest. But the cost of using lumber to build multiple acres of growing boxes for our expansion was prohibitive, dimensional lumber makes the boxes too heavy to move, and we wanted to create a model that was affordable, replicable, and movable.

We researched every possible alternative, from waxed watermelon boxes, to Dutch bulb boxes (of which there are millions), to custom-manufactured plastic bins. Eventually we settled on using pallet collars. These are thin plywood bottomless boxes that are placed on top of used shipping pallets. This approach was cost-effective and fulfilled our need to be able to move the growing boxes on short notice.

The Chinese company that produced these pallet collars came up with a price of $13 Canadian per unit and agreed to imprint our logo on the sides of each collar. The salesperson was aggressive, the language barrier made it challenging to negotiate, and the time difference between Vancouver and China made it difficult to connect. But in the end this seemed liked the best option, so we placed an order for forty-five hundred, enough to cover two farm sites.

My friend Jeff Wells was visiting from California the first day we installed those wooden boxes. It was a beautiful, sunny late-spring day, 2012, without any sign of the damp conditions that are common for much of the rest of the year. Jeff did not miss a beat, admonishing me for not painting or treating the plywood before it was installed.

Though I'd raised this concern and had even found a company to donate a nontoxic wood treatment for the plywood, we were already months late on getting started, and Seann persuaded me to move forward and install them as is. My gut knew this was a bad decision, but I let it go, not realizing at the time how costly this eventually would be, not just financially, but psychologically as well.

Within a year the plywood edges of the boxes were already beginning to come apart. In less than two years the boxes were fully delaminating and we were in a real crisis.

Most farms are in rural areas in locations that are private and invisible. Yet by design our farms are public and highly visible. Everything we do is seen by thousands of people from their apartment buildings and from the trains and walkways and roads that surround us.

Having our most basic infrastructure decaying in front of the world was embarrassing. Not only did it look bad, but it became impossible to till soil or plant or cultivate without breaking the boxes. Viewed from outside the fence it looked like we had given up. I had people approach me asking if Sole Food was "calling it quits" or "going out of business."

During an evening dinner held at our newly planted orchard, the city pulsing around us, several community leaders, including the mayor, stood up unprompted and spoke about our work and its importance for the city and its inhabitants.

When the mayor finished his talk, Terry Hui stood up and, in his humble way, raised concerns about the decaying wood boxes that graced the parking lot adjacent to his condominium sales office. He was excited about the challenge of designing and manufacturing new plastic growing boxes for us. And he pledged to take this project on and to pay for it.

I don't think at the time either Terry or his partner, Olivia, who oversaw the design and manufacturing, realized how much it would cost to

manufacture the boxes. Nor did they envision the design and manufacturing challenges that we would encounter. But like much associated with Sole Food, even with so much effort put into early planning, sometimes it's better not to know what you are getting into. Now we plant in new black plastic boxes, indestructible and designed for this purpose.

———————

Everyone from Sole Food who's come to visit Foxglove is acutely aware of the challenges of farming in a city like Vancouver, the high costs of everything, the value of every square inch, the developers racing to squeeze condos into every crack and crevice.

The neighborhood that is home to many of our staff is caught in a major transition. Even as we sit together during the workshop on my family farm, boardinghouses and old hotels in the Downtown Eastside are being torn down, replaced with condominiums, restaurants, and all the other trappings of a city in the throes of prosperity. I wonder, *Where will everyone go?* While I have never discussed this in depth with our staff, I know how much they are affected and I imagine it is the subject of private conversations among them.

The presence of the Sole Food farmers on Salt Spring Island reminds me of the reality of our farms in Vancouver and how they will have to adapt and flex as neighborhoods change. These questions of adaptability, however, I've come to see as the core questions any urban farmer must ask herself. As the community changes, what community are we serving, what kind of relationship do we have or want with that community, and is ours a farm that everyone can embrace and benefit from? What ethnicities and cultures are represented in our current neighborhoods? Who will be represented as they change? And a big one: What do they like to eat?

How will the farm's neighbors respond to people coming to work each day, visitors, occasional noise from machinery, a sales outlet, composting activities, and different smells? Will a farm in the neighborhood be well received, or will there be resistance?

The farm I developed in Goleta, California, in the early 1980s started life surrounded by other farms and open land. The farm was owned by a music professor named Roger Chapman and his wife, Cornelia, the daughter of the founder of Royal Dutch Shell oil company. I had a good relationship with the Chapmans. I was given the creative freedom to develop and manage that twelve and a half acres, and they got to watch as the land improved, orchards were established, and fields were planted.

They also witnessed the development of a huge market for the abundant fruits and vegetables we produced. We taught agricultural practices to a community that was, just at that moment, undergoing a rapid change from rural to urban, and even suburban. In time developers gobbled up the land around us and the farm became an island floating in a sea of tract homes and shopping centers. The new neighbors that moved in did not always appreciate the sounds of roosters crowing, the smell of compost, or the hustle and bustle of daily farm activities. They complained, and I was even threatened with jail time over activities that for years had been normal and were essential to the farm.

Eventually the land itself came under threat. Our acres had been zoned for fifty-two condominiums, and while the owners of the land were supportive and happy to see the success of the farm and how it had become a major landmark in the community, every indication was that their kids would eventually want to cash in.

The development pressures, the public challenges over the crows of our roosters, the making of compost, the signs advertising our fresh products, and the smells and sounds of a farm required that we fully engage with the new community of urbanites that surrounded us and that together we try to come to terms with how suburbia, growing up all around us, and a farm could get along.

Our situation was emblematic. The collision of rural and urban was playing out all over the country as the dollar value of the land—notable in the suburbs and cities alike—became far greater than the value of the food it produced.

Eventually we were successful in preserving that land, placing it under a conservation easement that preserved the farm forever. But not every farm or community has been so lucky. According to the American Farmland Trust, fifty acres of prime agricultural land is paved over and developed every hour in North America. Our success in California was based on the public education and community engagement we had been doing for years, well before the land came under threat. Having witnessed in California the hyperspeed of development of some of the best farmland in the world, the almost complete disconnection of most people to how their food comes to them, and the pressure the Chapmans were under to sell that land to the highest bidder, it was clear to me early on, even then, that we needed the community around that land to identify with it as if it were their own. I knew that the time would come when we would need community support, need the broader neighborhood to rise up in our defense—which, when asked, they did with their voices and with a million dollars in donations. That farm had a high profile in the media, it had a symbolic role in the broader community and the world, and its preservation would reverberate beyond the edges of its fields and orchards. I like to think of our workshops at Foxglove as continuation of the community-building activities that saved our California farm.

Of all the people in the room today participating in this workshop, most are young. If they are already farming they are leasing the land, as I once did and as Sole Food does in Vancouver, and if they plan to farm in the city they will never be able to service a mortgage from farming income. Few people in Vancouver could ever afford to own the land they are farming on. Sole Food certainly cannot.

"But ownership as the dominant model for young and beginning farmers is highly overrated," I suggest. Then I continue: "Long-term leases for open-field rural farms are a better way to get started, if those leases are for a minimum of ten years. It normally takes half that amount of time just to get to know land; to improve the soils; to understand how the light and air move; to start to become intimate with the biology, the broader ecology, the human and natural communities within and around the land; and to establish stability in a new farming operation."

And yet what I also know is that for those who wish to farm commercially in the city, ten-year time frames are unrealistic and often impossible. Urban land is too valuable for landowners to tie up in any more than a three-year lease. And so we seek out those situations that have potential for longer tenure, but we settle for less.

At Sole Food we've learned that finding a site with sympathetic ownership is important. Farm use on urban land is not a typical use and as such requires a landowner who is open-minded, willing to consider new ideas.

I tell the group that building a relationship with the landowner is essential. Leases are merely the documents that present a set of legal terms; relationships secure those terms and provide a supportive environment for the work. I reflect on how lucky we have been with our relationships with those who own the land we farm—the city of Vancouver, Concord Pacific, and the owners of the Astoria.

"So stay in touch with your landlord," I tell the group. "Provide them with the food that is growing on their property, educate them, and work to make their land more beautiful than you found it."

As we prepare for a break I consider the big ideas I've been presenting. I imagine that for many of those who are just getting started it can be a lot to digest. While looking back over the years, I can tend to make our work seem easier than it ever has been. And like many of these young people, and like many of the Sole Food farmers, I was impatient in my youth, sought out shortcuts, did not always trust the process that

Ownership: An Overrated Model

I spent my first thirty years farming on land that I did not own. It's not that I did not think about ownership or desire it; I simply had my head down working and did not take the time to pursue it.

I farmed in Southern California, and I remember my father telling me, "Michael, you really should buy a house in Santa Barbara," the closest big town to the farm.

"Dad," I said, "those houses are selling for $150,000!" I thought this was a staggering sum at the time, and so I did not buy. Less than thirty years later, that $150,000 house is worth $1.5 million.

The truth is that I don't like taking financial risks. Kind of crazy when you consider that every seed or tree I plant begins a lifetime of risk the moment it is placed in the ground.

If you grew up in America as I did, ownership is bred into your cells, part of the national religion.

I was almost fifty before I joined the ranks of "owners." I was lucky to buy land in another country at a time when the exchange rate was favorable and the real estate prices were a fraction of those in California. I appealed to the local bank, which, despite my having no credit history, provided me with a mortgage. It is remarkable that those of us who've never accumulated debt are considered a credit risk!

For a long time on the farm I own I felt like I was the manager, a role I was very familiar with, waiting for the real owner to show up. But when the roof needed repair or the plumbing failed the responsibility was mine: either do the work or pay for it to be done. The bills arrived continuously—for insurance, mortgage payments, property taxes. I soon realized that the freedom

required plodding and planning and taking my time. "Agriculture is a slow and patient livelihood," I say as everyone files out of the room. "It requires more than a good idea or the right clothes or tools or reading the right books." Or even attending a workshop, I think.

———

"What is the most important agricultural skill?" I query the group as we settle back in for a final session at Foxglove.

"Soil science," Rob calls out.

"Mechanics."

"Entomology."

and autonomy that I thought came with holding title was an illusion.

Not everyone can participate in the dream of ownership, but despite my relative privilege to do so, I concluded that it is a highly overrated model. Often I run into someone who says, "You inspired me to buy a farm." I'm never quite sure whether to offer my congratulations or my condolences.

Knowledge and experience are essential assets for a farmer, but they can only come with time. Land is a necessity from the start, but ownership of that land is not. There is a lot of land out there that is privately owned by people who would like to see it farmed but do not have the skills or the desire to do it themselves. Creating long-term leases that address all the potential challenges that will arise over time is, in my mind, a better model in these times when few can service a mortgage from farm income.

I know all too well that farmers make a significant and often unseen investment in the land they occupy. Society does not place dollar values on improving soils or slowing runoff or planting trees. Farmers must be able to stay on leased land long enough to realize a return on all of those less visible things they do to make land productive.

But I do not believe that the challenge new farmers face is access to land. I believe it is access to the financial and knowledge capital to develop and farm that land. We may need to look deeply at our need to "own," try to understand why it is so seductive, and come up with creative alternatives. The responsibility of our food and how it comes to us, the responsibility of how we use and steward our land, should not belong solely to the 2 percent of the population we call farmers. That responsibility belongs to us all.

"Marketing," Kenny says.

Others in the group think it is important to know how to operate a tractor, do crop planning, lay out a new field, make compost.

"All these things are essential," I say. "You're right. But the most important skill, I believe, is the art of observation."

I use the word *art* as much as I can in describing the work that we do. I'm trying to break down stereotypes and open eyes to the depth of what we do as farmers. It surprises and sometimes irks me that people still have a hard time seeing agriculture as an art. Beneath all the hoopla around food these days remains the underlying perception that farming is a mechanical or industrial job, and that unlike other professions it does not require discipline or extensive training or study or education.

So I insist that few professions call on the range and complexity of skills that farming does: mechanics, carpentry, biology, botany, soil science, refrigeration, electrical, welding, sales and marketing, public relations, retail display, and on and on.

But beyond these skills, the thing that really defines good art—and farming artistically—is observation, really seeing the world, editing elements from that world, and interpreting those observations into a melody, a poem, a painting, or a well-made box of beautiful and flavorful food.

They all know I had no formal training as a farmer, did not go to college. I have not read many books, and when I was first getting interested in agriculture there were few apprenticeships available. We all have different ways of learning and doing, and my way has always been to jump in and figure things out as I go. I explain that nature is a wonderful teacher if you are paying attention, and as a young man I relished my daily walks, took careful notes on what I was seeing, and eventually developed the ability to make concrete decisions and lists of tasks from my observations. I still take these walks.

"My walks are intentional," I say, "entirely devoted to understanding what's going on within and around my farm." And doing these walks every week, year after year, has opened me up to the subtle, wonderful, and magical transformations that are constantly taking place.

I'm clear with the group that when I'm on a walk I leave the electronic devices behind and, again, make use of an old-fashioned notebook and pen.

"I like to have a notebook that fits in my back pocket," I say. "I walk the same route every time, and I do it two to three times a week. Each time is different, sometimes subtly so, while other times there are dramatic changes even from one day to the next." For me, the same route is important because I want to see things in a continuum; I want to have a visual baseline to be able to recognize changes.

"At first," I continue, "walking the farms can seem like a difficult process, because not everything is revealed in the beginning. Nature wants to know that we are committed, not just passing through, before it reveals itself. So when I first started this practice, the first month or so, I would come away feeling stupid, my notes reflecting simple identifications like 'tree' or 'cow' or 'blade of grass.'" When I say this—and I say it often—I know it sounds lame, but it took time for me to see what the land I farm had to say.

"When I carried out this practice enough, though, and when I became really present and paid attention, when I was quiet and listening, my notebook filled up with detailed graphic descriptions, new revelations, ideas, and tasks to be done."

I tell the group about when I first discovered the dynamic transition from dormancy to full bloom my first season in the peach orchard. Over a couple of weeks I witnessed this incredible drama unfolding from the swelling of the buds to the popcorn stage of the flowers, from full glorious bloom to fruit. I was witnessing the massive, simultaneous birth of thousands of little fuzzy peach babies, each one enveloped in jackets of fading flowers that I could slide off with my fingers, revealing the fruit, newly formed.

As a young farmer, these moments sustained me through the long days and hard work and helped me to recognize my part in that dynamic, ever-changing world. "I was learning farming aikido," I say. "I came to understand when to insert myself, gently nudging things along, and when to stand back, observe, and allow things to take their course."

With proper attention, I feel that I can enter the slipstream of biological life on the farm without having to manage or control it. To do this requires careful observation, and nothing has served me as well as developing this skill. Of all the farming techniques I've practiced, of the few books I've read, of all of the courses I've attended or apprenticeships I've taken, none are worth anything unless they serve a deeper observational understanding of the complex dynamics that are constantly playing out within my farm.

"But farms," I say, "are imposed environments, and we must take an active role by providing the conditions for health and productivity. Only then can we witness the reactions and the interactions that occur as a result of what we do. This is how the farm tells us what it needs. Spraying the orchard, for example, with a seaweed extract and seeing within a couple of weeks the dramatic change in leaf color and plant vitality. Or cultivating carrots and, immediately after, witnessing carrot tops standing tall and straight, as if they could now—with some attention—breathe deeply and grow fully in ground that's well aerated and cleared of competition."

As I refined this ability to see the farm, I began to notice more and more subtlety. The farm speaks to me now in more interesting and complicated ways than it ever has before.

I remind everyone that to be a good observer, you must engage your whole being, all of your senses—smelling, seeing, feeling, and, of course, eating.

By this point in the day, the group has been sitting for a while, and I can feel that some of them are growing restless. My mention of eating is making it worse.

Beginner's Mind:
Walking, Seeing, and Responding

Walking the farm is our most important job. Planning satisfies our need to be organized, to think ahead, and to avoid reactive management, but it is the walks that provide real-time, bio-appropriate, moment-by-moment information that informs our daily decisions. Walking keeps us humble, reminds us of our place in the broader system, and brings us into intimate contact with the real world.

Biological systems never stay the same. We can plot and plan all we want, but in the end nature is always changing. To become a part of the slipstream of our farms, to learn to respond rather than control, to be so flexible that we can bend and stretch and change at any moment, this is the core challenge of our work. Our best work comes when we approach our farms with a beginner's mind: no preconceptions, no fixed ideas or plans, always open to the ever-changing moment.

Walk often, stay open, bring a notebook and a pen.

I decide to close this session. But I also know I need to emphasize that there is a difference between *looking* and *seeing* and that the writing we do in support of seeing the land must be more than a sweet paean. To do this, we must convert observations into a real work list. I don't hear voices exactly, but the plants and soil and life in and around the farm do tell me what to do.

We stop to eat. Eggplant, basil, and tomato sandwiches. There is iced mint tea, a big salad of mixed lettuces, arugula, and radicchio, and one of our yellow-fleshed watermelons that everyone has been eyeing for dessert. We eat outside at picnic tables shaded by the old Transparent apple tree that John Rogers planted, some of us sprawled in the sun on the grass, others off on their own.

It thrills me to see my family farm used this way, to know that our outreach is multidimensional. Improving our soils and growing good food is our foundation, but to be able to use this land to teach and inspire and as a place for people to retreat and restore is fantastic.

And feeling this satisfaction, I am reminded that this meal, harvested from fields visible from where we sit, is as powerful a teaching tool as anything that has been said or demonstrated all morning. It's the edible embodiment of all of the elements—the knowledge and passion of the farmers, the energy of the land, and the millions of tiny organisms that inhabit our soils. No list, no plan, no lease, no history lesson, can match this. Nothing matters on a farm as much as the food. And by the end of this meal, that's the lesson contained in each and every face.

CHAPTER 5

Boxes and Bellies

t's mid-February and I'm in a Vancouver, British Columbia, restaurant I shall not identify by name. The menu tells me, "The majority of our seasonal ingredients are sourced from local, sustainable, and organic producers."

There's a tomato and mozzarella salad, a Niçoise salad with "fresh" green beans, and a Moroccan-style dish featuring a medley of summer squash, eggplant, and peppers. And so I pop (as I always do) the inevitable question to the poor soul who got stuck waiting on my table: "Where exactly did you source these fresh summer vegetables in February, in Canada, with three and a half months of cold weather behind us and most farms put to bed long ago, not to wake up again until March or April?" My family members and friends always risk some level of embarrassment when they go out to eat with me.

I don't think it's always intentional, but menu fraud is a fact of life in these times of *local, seasonal,* and *sustainable.*

I know it's one hell of a challenge to run a restaurant; most of them don't survive their first year. Given those pressures, I also understand the desire to use the language that everyone wants to hear. My hat goes off to those who are willing to brave the long and late hours, the unrelenting demands to perform and be consistent, the need to always be on, and the unbelievable financial challenges and risks.

But I also cringe when I see the "pre-cut" truck arrive at a restaurant near where I live, disgorging boxes filled with already chunked and sliced and grated bagged "vegetables," which are carted through the door directly under the sign that prominently announces FARM-TO-TABLE.

I'm aware too of how much of the praise that chefs shower on growers is lip service, and how few chefs have a rich depth of commitment to the source, the foundation, the people and the places where their ingredients come from. I know for certain that all the culinary skill in the world, all the sophisticated training and experience, all the complex sauces and preparations imaginable cannot possibly bring to life ingredients that are already dead on arrival to the restaurant kitchen.

At Sole Food, our small-scale intensive model of farming requires that we grow some high-value specialty products to pay the bills, a majority of which end up in the kitchens of many of Vancouver's most notable chefs. Sole Food sells to more than thirty restaurants in the city. These restaurant relationships have become a cornerstone in how we market our products, and our relationships with chefs inform our crop planning and their menu planning.

It may seem crazy to have to deal with a bunch of upstart independent farmers in order to pull together a reliable local menu, but I know that we are absolutely essential to the success of the chefs we supply. Yes, chefs: Sing our praises loud and clear, but only do so if you mean it.

This thing we now call a food system—*is food really a "system"?*—is about relationships: interpersonal, ecological, biological, social, and political. Each one of those relationships has to be nurtured to function well. I will bend over backward to take care of a chef whose commitment is real and true and whose loyalty extends beyond a single order.

Navigating through the complex needs of eaters these days is incredibly difficult. I wonder how restaurants can possibly respond to the range of diets and food sensitivities their customers may have. I look forward to the day when people expand their thinking beyond gluten-free, vegan, omnivore, locavore, pescatarian, and vegetarian and inquire instead—or additionally—about whether the farmer and his family are well paid, the land has been well cared for, and the cook was in a happy mood when he or she prepared the meal.

I ask my friend, chef Dave Gunawan—co-owner of Farmer's Apprentice restaurant—why he bothers going through all the hassle to buy from smaller local producers, like us. Dave is pretty mellow and even-keeled, nearly impossible to get riled up. "Buying from those small farms is more expensive and more time consuming," I remind him, trying to elicit a response. "Why not do one-stop shopping, make that one and only phone call to a distributor, order off a list, and have everything magically arrive at the door of your kitchen the same afternoon?"

Dave has an answer: "It's frustrating to be involved with food but to be powerless about the ingredients. What we are trying to do is a political act. It's a way to be different, to be more conscious. It's a way to fight against industrialization. And of course the quality is a natural gain."

Clearly, he has to be concerned with the economics, but they work backward, Dave says. "The economic structure has to fit the goals and the amazing food, not the other way around. We have more control over the service and labor than we do over the ingredients. In the end it's not so much a business model as it is a philosophical model." Dave tells me that while they list the variety names of the products they use—kabocha squash, Bosc pears, Mutsu and Ambrosia apples, and so forth—on their menus, they avoid listing the names of their suppliers or making any grand statements. "We're not here to tell you something. It's not a marketing strategy. This is just what we do."

Dave describes for me his learning curve, how he's gotten to meet farmers, and how at the restaurant everyone gains an understanding of the ingredients. "We wanted to get into the movement, but initially we didn't know why, so we went to meet the farmers to understand what they were doing. We've learned so much in these last years. We didn't know anything about pigs and how they were produced, or how cold weather improves the quality of brussels sprouts. All this rich information can only come from building a relationship."

Dave and I first met through mutual friends over dinner at my farm on Salt Spring Island. In 2013 he opened the Farmer's Apprentice and from day one has been a loyal supporter of Sole Food, using well over a ton of our products since the restaurant opened. But Dave also saw right away that we take the social element to a whole different level. And so he began inviting our farmers to eat at his restaurant every week, at no cost. Dave says, "We wanted these people to know how the produce they are growing and harvesting is being used and treated." Some of our farmers—Kenny, for one—have told Dave that what they eat at his restaurant is the best food they've ever had in their lives. He senses that allowing our team to see that transformation from the field to the delivery truck to the table inspires them and offers them an understanding of the cycle and the knowledge that their work is being made even more meaningful through what gets served in restaurants like his around town.

As Dave sees it, "The folks from the Downtown Eastside are normally kept at a distance from good food and how to gather around it." These weekly meals allow the farmers at Sole Food, myself included, to see the bigger picture, the end product.

Our staff are totally unfamiliar with the kinds of dishes Dave and his team prepare. These chefs are adventurous and experimental, sometimes going beyond what my own palate can understand. I doubt that any of our farmers have even heard of Ossau-Iraty cheese, or Coronation grapes, smoked cabbage, sea buckthorn, or crosnes. And now they're being invited to sit down at a linen-covered table and have someone serve those items to them. This weekly offering, a seat at the table every Sunday for a few of our co-workers, has been an awakening for some. But I've also watched how uncomfortable our farmers can be in a restaurant setting. Occasionally—beyond these weekly meals—we've scheduled meetings at some of the restaurants we supply, and I've observed the unease, the late arrivals, the desire to go outside and have a smoke, when seated at a table where someone is cooking for them and serving them. These restaurants are, for many people, a foreign world that those of us with privilege take for granted.

The chefs we work with appreciate our farmers and the hard work they do, work that may never enter the consciousness of restaurant patrons who may see our world as strange and even intimidating. Dave describes "the meticulous nature" of the products we deliver from Sole Food. The French Breakfast radish arrives with its tops on, the beets in multicolored array; the chicories are vibrant and varied. Everything is packed and presented with incredible care and professionalism. We deliver a bounty. And I feel a sense of pride when I hear Dave describe our produce, not for myself, but on behalf of everyone on our team who has made all this happen.

Although it is too costly to do regularly, I try to patronize the restaurants we sell to—Dave's included. I want to know more about their menus, how they use the products they get from us. I want to get to know them within their world, and for them to see us eating their food and know that Sole Food believes we're involved in an exchange. And while it is expensive, it closes the circle, provides us all—farmer and chef alike—with the insight and inspiration we need to grow and cook the ways we do. But I sometimes wonder what role high-priced fine-dining restaurants play in our society. I have seen how well-known chefs and the influential clientele they serve can elevate the public dialogue about food and farming. I have seen the attention that media gives to those whose artistry turns good ingredients into high art. But I also wonder whether these temples of food worship are just there to keep rich folks happy or whether they can actually be a force for change in the world.

Occasionally we field complaints from chefs about the cost of the products we supply. And while I notice that $1 worth of Sole Food arugula or salad might appear on a menu at a price of $8 or $10, I understand the unique financial challenges that chefs face. Too many are caught between their personal commitment to using the best ingredients and

Birth and Delivery (aka Harvest)

It's easy to assume that skill is required to grow food well, but that harvesting and preparing food for sale is somehow self-evident and without subtlety or detail in its execution.

I'm not sure about other people's farms, but I have yet to have a spontaneous harvest experience where I show up on a harvest morning to find that all the lettuce has perfectly cut and washed itself and fallen into the boxes, or that peppers have hopped off the plants and lined themselves up in crates, or that all the perfectly ripe melons magically rolled out of the field and into the truck.

Every crop requires a particular, variety-specific set of harvest judgments and techniques that make all the difference between perfect texture, flavor, and cosmetic quality—and not. It is this stage in the process that requires the most careful and watchful attention. What color is a ripe strawberry? Is it red, or is it actually crimson? What is true for an Albion strawberry is not true for a Mara des Bois. Every variety is subtly different. Making your way through a field quickly, moving leaves aside, lifting and looking at the underside of every fruit, involves making instantaneous choices on which to leave on the stem and which to pick and put in the box. An experienced strawberry picker has, by tasting and looking, developed special neurons that connect eye, palate, and hand to inform a million little decisions per second.

the demands of restaurant owners who require that they keep their food costs down. It is difficult when your bonus is tied to whether you are able to keep your costs low, and you're not getting a high salary to begin with.

The financial challenges for both farmers and chefs require a constant balancing act. The cost of growing food well is high, and I suspect if growers did a thorough cost analysis of each product they grow, if they really knew what it costs to prepare soil, seed, cultivate, irrigate, cultivate again, harvest, wash, pack, and sell, they would be surprised at how little per hour they earn. It is difficult to explain this to anyone on the buying end of this relationship, especially in the few brief moments we have together and with the knowledge that they see similar products available cheap from the fields of industrial agriculture.

And if it's difficult to explain quickly why what we produce is more expensive, it's downright impossible to explain that cheap food is the result of massive subsidies, soil loss and degradation, groundwater depletion and pollution, and farmworkers who are not always well cared

I have always offered a "melon back" guarantee. I sleep better knowing that if someone spends $5 or $6 for one of our French melons or $8 or $9 for a watermelon and it is not fully a religious experience, they can get a replacement.

Growing melons well requires refined skill, especially in our northwestern climate, but harvesting requires almost psychic abilities to get it right. French melons, unlike a more common western cantaloupe that is ready when it "slips" off the vine, are harvested primarily by smell, with the support of some key visual cues. It is quite entertaining for visitors during melon season to witness us crawling around with our noses to the ground sniffing out ripe Charentais or Petit Gris or Cavaillon melons.

Watermelons require a different set of instincts and skills. Without any fragrance, watermelons require the farmer to develop a sixth sense to harvest them successfully. Yes, we check the underside of the melon that has been sitting on the ground for a change of color from green to white, the drying-out of the tendril closest to the melon, the drying and dying of the primary leaf, but tapping and listening for that perfect deep tonal sound guides our final decision. No one seems to entirely agree on approach here, and so I suspect that in the end it is more feeling, instinct, and confidence based on experience that moves us in our decision to break a watermelon free from its mother vine or leave it for another day.

for and protected. These concerns are always on my mind. But there's some silver lining on this gray cloud. Because the battle we wage against cheap, industrial food also involves the essential quality I think chefs value most—taste. Sole Food products, food grown well, are dramatically more flavorful than any product of an industrial factory system.

During those times when Sole Food's products are not available I provide Farmer's Apprentice with other food grown well, delivered from Foxglove Farm. When I pull up to the sidewalk in front of the restaurant to unload an order of fresh lima beans, fingerling potatoes, sweet onions, roasted peppers, watermelon, and winter squash, Dave and a couple members of his crew greet me at the sidewalk. There are hugs. We then lift lids off boxes, and Dave and the others inspect the products. I feel a palpable sense of excitement swirling around as the possibilities that these products will provide are revealed to these young, adventurous, and super-creative chefs. I am usually invited in for something to eat, or, if I am in a hurry, I'm handed a container of soup or some other treat for the road. As I leave, I jokingly remind Dave not to bury or hide my vegetables in one of his beautiful creations, to do right by them, to make them stand out.

––––––––

A selfie posted online of Kelsey Brick standing inside one of our farms caught our attention. She had applied for a marketing position with us in September 2013. She really wanted the job, and, in the end, her stunt worked.

The marketing job would be key for us, and there was a lot of time spent around hiring for it. The position was responsible for processing all the food as it came from the fields, packing the orders, networking with chefs, and handling sales. She would be our quality control and our face to the wholesale and restaurant world.

At first we had a fixed idea about who we were looking for. The perfect candidate would have extensive marketing experience and know produce and chefs and the wholesale trade inside and out. But the truth is that no one was ever going to live up to that.

Kelsey had some of those skills, but her enthusiasm and stick-with-it energy was where she shone. I liked that she had experience in restaurants and that she understood that culture. Kelsey had been working in Alberta for a company that provided food service to the lodges and camps that housed and fed the people working in the oil fields. What stood out to her there, she's told me, was that the people she was cooking with often knew more about the knives they were using than the food they were cutting. She started working in kitchens at the age of sixteen, studied professional baking and pastry arts, and cooked in fine-dining restaurants, bakeries, and pubs. When she applied to Sole Food, Kelsey had returned from her stint in Alberta and been living in Vancouver for a couple of months. When she left the food service company, she didn't want to leave food entirely, but she tells me she was feeling bitter and looking for a way to start working toward connecting culinary professionals like herself with the food they were using.

She remembers the difficulties she faced when she first started the job in spring 2014. For one, despite the selfie stunt, it took us a long time to hire her in the first place. "You made me wait," she recalls. And she was right. Seann and I wanted to make the right choice; we had hired the wrong person for this position previously, and there was a lot of debate about whom to bring on. Once she started work, she remembers, "there was a lot going on trying to learn a new job, and I was thrown into the deep end of it. It definitely pushed my limits to stay on top of things, and there was a point that I thought I was drowning." Even so, with Kelsey it became clear very quickly that we had finally hired the right person.

While we have since built a new processing and packing area at our farm on the Concord Pacific land, when Kelsey started working with us the post-harvest processing area consisted of two 10-by-10-foot portable tents set up against a shipping container in the middle of a parking lot. Inside were a freestanding stainless sink, a washing machine (used on the spin cycle to dry salad mix), and a couple of tables. Now imagine all the produce from four acres of food production funneling through that makeshift, temporary facility several times a week. It gets hot on that pavement. There were dozens of people coming and going, dropping off newly harvested product from several sites, thirty chefs all expecting on-time deliveries of perfect food, produce for farmers markets going out several times a week, and a weekly community supported agriculture food-share program to supply. On top of that, the post-harvest washing facility sat smack in the middle of all the farm traffic, with staff coming and going. It was also the first thing visitors encountered, and there are lots of visitors.

People are always wandering into the farm wanting to know what's going on or to buy veggies and have a chat. And because the processing area was up front and most visible Kelsey became a de facto greeter. Normally, she'd tell these visitors to show themselves around the farm. Still today they often ask if we are a community garden and want to know if they can buy soil. These people are surprised and sometimes concerned when Kelsey responds that they cannot buy even two cups of it. We need our soil.

Film crews visit Sole Food too. They use the neighboring lots for staging. There are also truck and bus drivers, security personnel, and people attending music and sports events who use the parking lots that surround the farm. Kelsey fields questions about parking fees, which she has no answer for, and there are those who very sheepishly and desperately ask if they can use the portable toilet. But then there are those people who want to tell Kelsey stories about their parents or grandparents' gardens or their own gardens. As Kelsey has told me, with typical good humor, "I'm the worst person to be asked about how to grow anything. I pack vegetables, I don't grow them. I have one green onion in my garden and the bugs ate all my kale . . . such jerks."

Some of Kelsey's strongest relationships in Vancouver are with local chefs. Before she started on the farm we asked her to spend a week visiting as many of the chefs we sell to as she could. In that week she visited a dozen restaurants, and in return she invited a dozen chefs down to the farms to see what we do. These visits continue today.

The chefs are impressed and amazed by how close we are to the dining rooms where they serve the food. Kelsey calls this "hyper local." And there is a method to Kelsey's madness, as it were, in bringing these chefs around. "I have a theory," she tells me, "that the more educated chefs are, the more willing they are to work with you. The hardest people to sell to are the people who have very limited knowledge. Being able to get chefs to come to the farm and see what we're doing is essential."

I couldn't agree more. It's one of the great lessons of my life.

During the thirty years that I farmed in California we developed a lot of close relationships with chefs. James Beard Award winner Mike Tusk and his wife, Lindsay, who own and operate Quince restaurant in San Francisco, were loyal supporters of the products that we grew. The late Judy Rodgers of Zuni Café came to the farm on many occasions to teach our Field to the Plate workshops. She and I would lead a farm tour and talk about what we were seeing from both the culinary and the agricultural perspective. She had such natural intelligence about food and carried herself with little of the attitude that often comes with the celebrity-chef mantle.

Alice Waters and her family spent a number of post-Christmas holidays with us on the farm, where we cooked from the fields while dreaming up ideas for new programs and projects. We supplied white asparagus, clementine mandarins, baby artichokes, and French filet beans to Chez Panisse during those years, shipping the produce by way of the Greyhound bus to Oakland, which left only a quick schlep to Berkeley. The eleven-acre farm I started at the Midland School, a college-prep boarding school in Los Olivos, California, was one of the early inspirations for Waters's Edible Schoolyard Project.

Over the years some of those restaurants sent staff members to work with us for a few days, and it was obvious to me how that brief exposure to the workings of the farm informed the relationship between the chefs and the ingredients they were using. It put a face to the food and gave them a new language. Culinary school and working in restaurant kitchens do not teach young chefs how much attention to detail is required to artfully harvest a bunch of beets or carrots or those perfect little haricot vert beans. They did not know the contortions we went through to produce white asparagus or how many years it took for the clementine mandarin trees to mature before the first harvest. A day or two of work in our fields was all it took to dramatically raise the level of appreciation for those boxes of beautiful food that had been magically appearing at the back doors of their kitchens.

I treasured the wonderful generosity in those early farmer–chef relationships. We were all discovering and redefining our connection with food. Those early days were the beginnings of a culinary revolution that has now rooted itself deeply. They were heady times, and those exchanges between the agricultural and the culinary inspired new crop and variety choices, and a total awakening in new growing, harvesting, and cooking ideas.

Those grazing tours I did with chefs provided us with ideas and inspiration that we took back to our fields and kitchens. My experiments in growing white asparagus, which became one of my signature products, started with the discovery I made with a chef of a random white spear that had been mistakenly covered by a piece of black plastic. Growing baby artichokes—instead of the large, less tender ones that were so common at that time—was inspired by conversations around the table at Zuni Café. When chefs found out about the volume of food that we rejected for cosmetic reasons, or how often a bumper crop of tomatoes or peppers could not be sold, they opened up their kitchens on days when the restaurant was closed and processed those products for later use.

And of course some wonderful friendships were established, forged over long, delicious meals together. Many of them continue to this day.

Sole Food products are literally grown down the street from the restaurants and markets we supply. As Kelsey says, our version of local is *hyper*. No need for the Greyhound bus.

The crownlike circular steel sports stadium BC Place hovers over our largest farm on False Creek. The stadium is home to both the regional soccer and football teams, it hosted the opening and closing ceremonies for the 2010 Winter Olympic games, and Michael Jackson, Madonna, and Paul McCartney have all performed there. Indeed, the food that McCartney was served when he was at BC Place was grown by Sole Food on the parking lot visible from the windows of the stadium. In an article in the *Vancouver Sun* about this, the reporter noted that the food was grown by "people with broken wings who learned to fly."

Though their orders meet our minimum dollar value for delivery, the chefs from Center Plate restaurant, located in the stadium, will roll a cart several blocks down Pacific Avenue to the farm to pick up fresh ingredients. Kelsey has told me that for some of these chefs, excursions to Sole Food provide an escape. The kitchen, she says, "is a high-pressure zone. It's nice to be able to get out, like a mini field trip." Often chefs will come down and walk around and notice products that they haven't ever purchased. Kelsey will pick a leaf of mustard or chicory or a radish. She tells me that the chefs love that, and she enjoys being able to connect with chefs this way. She recalled a day when a Vancouver chef named Shingo came down to collect baby Hakurei turnips and French Breakfast radishes. While he was on the farm, Kelsey gave him a tour and treated him to some samples of items he was not familiar with. She knows well that it's this kind of experience, coupled with her commitment to learning about the very specific needs of the Japanese restaurant Shingo works for, that led to his loyalty to Sole Food.

———————

There are people in this world who have been through hell and back and still seem to come out the other side with their heads held high and a smile on their faces. It's not that the suffering is not there, it's just that they've chosen not to let it define who they are.

Years of hard-core use of booze, heroin, and crack. Having his large intestine removed. Chronic emphysema. Long-term poverty. Alain has had more than his share of hard times.

While there is a danger here in projecting or idealizing, it is safe to say that Alain is one of those people who has learned to accept and roll with pain and suffering, to not let it sink him. My own life's challenges pale in comparison with those I see among Sole Food's crew. I'm not suggesting any form of perfection here. Alain has baggage like the rest of us, and he's leaned on some pretty heavy stuff to get through his pain in the past. It's just that he's managed to recover, rise above it, and raise three kids—a stepson, Dion; a stepdaughter, Destiny; and a son of his own, Jordan. He's maintained his good attitude and become one of the star members of our farm team. This means a lot to him and to us. Alain's success is our success as an organization as well.

Alain

"You never lose addiction," Alain tells me. "It's like a gorilla is on your back doing push-ups all the time. Some stay away from it for years and then go right back in. Crack is like having an orgasm in your head, but it lasts a lot longer. If the devil lives in anything it would be in crack. It becomes your god."

Alain, now fifty-seven, and I are the oldest members on the farm team. By contrast, Kelsey is almost thirty years younger than we are. While I only have a few years on him, Alain loves to give me a hard time about my age. I laugh, but I also cringe, not so much because of my personal issues around getting old, but more because I got the lucky genetics, and I look so much younger than he does. Most of our crew has lived pretty hard, put their bodies through stuff that I could never have survived. It's not difficult to see what kind of life people have had by looking into their eyes or at the lines in their face, seeing their posture or how they walk. We all carry our life experiences, our grief, and our hardship around in every cell of our bodies.

Before coming to Sole Food—and also bringing his two sons—Alain had spent seven years working at United We Can sorting bottles. One day over lunch he reminisced: "I knew about the farm job, but I had no interest in farming. The only reason I came to Sole Food was because I thought my son Dion could do this work. I didn't want him working at the bottle depot. You walk outside the door of that place into chaos and temptation. I liked that job because there was no responsibility, and you get paid on the spot."

Dion has fetal alcohol syndrome. He's in the body of a young man— he last worked for Sole Food when he was eighteen—but his mind is like a child's. And Alain is caring for him completely on his own. Dion is strong, really strong, and if he has alcohol he can become violent. Alain looks after him with such patience, and I suspect he will be caring for him like a child for the rest of his life. At Sole Food, Dion turned and raked beds—"simple stuff," Alain told me.

"Dion has caused me more tears than anybody," Alain has said. And there's blame to go around. "I have a lot of pent-up resentment about what his mom did to him. I'd see her in the bars getting drunk with her big pregnant belly."

There's a contemporary mystique and some romantic ideas that are drawing well-educated, young white kids to farming. But I often wonder what it is that attracts people from the Downtown Eastside neighborhood—all poor, only moderately educated—to want to do this work. Is it just the paycheck, or is it something else?

Alain has told me that Sole Food revealed to him the difference between an occupation and a job. That afternoon at lunch he explained, "I now know I want to be a farmer. If you guys would have fired me in the first few years I would have cared less. Now I'd be hurt. I got into this for my son Dion, but now it's for me. I got exactly what I wanted. Seann and you don't give up on people." It's true—we try not to.

When Alain started, he really just wanted to dig and to be left alone. He's told me, though, that in time he came to learn how to bunch vegetables properly and how to handle people. He wasn't expecting to have to take on responsibility; he was shocked when he got a promotion—from farmworker to supervisor. That evening he went home and cried. "I got what I wanted," he told me. "I like learning from the best."

Farming has opened Alain's eyes in new ways. The world itself is surprising. "At first," he said, "every time we seeded I could not believe that the seeds actually came out of the ground. I always hope this shit grows. It's a lot of responsibility. It's got to be done right. Farming is the scariest job in the world for me, but to do this in the city is friggin' awesome." It's a scary job because, again, for Alain, it's become something far more—an occupation. A life.

At Sole Food, Alain has come to see another kind of difference too. It's not just job versus occupation. There's more: "When you live in this area around a lot of dope addicts, and then you meet people who have something going on and are doing something with their lives, it's like anything is possible. I love the concept of helping the people who need it the most."

When I was filling the boxes with soil for our new orchard, there were months of preparation to get to the place where we could plant the first tree. I wanted Alain to help me out. When it was time to plant, prune, tie, weed—anything that required someone to really go at it and pay attention—Alain was the one I wanted by my side. I know how much others at the farm had come to depend on him too, and also that it generated a little tension when I requested Alain's help. But we always worked it out. And Alain has become an excellent team member, a leader, and someone we can depend on.

Alain will say that we are saving people at Sole Food. The farm is a place to go, somewhere to start over. And Sole Food has always done the more difficult thing, he says. "It would have been easier to hire a bunch of people who know what they are doing. But we're helping a lot of people." As Alain tells me, "I always made excuses, I wasn't emotionally stable, but this job will hold on to you until you're ready to take the next step." Now he has an occupation.

Living and working on the Downtown Eastside there are many reminders of the life Alain has put behind him. I'm sure this must feed his compassion for those who are still struggling. "Drugs addicts are people, too," he's told me. "We all have feelings, we have kids, and we have families."

I often puzzled how it was that Alain's other son, Jordan, who grew up around so much craziness, turned out to be clean and sober, and so together. He was raised amid incredible violence, with police around all the time, people smoking crack in front of him. Alain was a single parent for Jordan's first five years.

So I asked Jordan—whose classic First Nation looks, jet-black hair, bronze-brown skin, and sweet smile endear him to everyone—how he came through all that and was able to stay clean and function so well in the world.

Alive and Still Moving

There is nothing static in nature: no fixed points, nothing locked in time and space. Everything is moving, changing, photosynthesizing, reproducing, growing, ripening, decaying, rotting, and transforming.

And while it is safe to say that at the point of harvest an apple or peach or carrot or tomato will contain only what the mother plant provided, it is also true that fruits or vegetables are zombie-like in their ability to continue to metabolize; they remain alive, evolving, breathing, and changing well after they have been plucked or pulled.

Months of planning, preparation, and hard work at Sole Food eventually result in beautiful food in boxes or bins or bags. But to complete the process and ensure that those foods arrive at their destinations still intact and delicious requires careful attention to the very particular and individual needs of each fruit or vegetable.

What we do in in the hours following harvest determines whether all the life and flavor and goodness in that food will actually be transferred to those who will consume it. Sadly, most retail produce departments are like morgues, their workers like undertakers, trying to create the appearance of life where it no longer exists.

Good farmers are like EMTs, caring for and transporting fresh living products at high speed like living organs waiting for transplant. Once pulled from the ground or severed from their mother plants, fruits and vegetables begin their march to their twilight, and there is a narrow window between optimal flavor and appearance and the onset of decay

"The reason I stayed away from all that stuff is that I saw addiction at its worst," he told me. "When I thought about trying Ecstasy, or weed, or other stuff, it brought me back to my childhood and I didn't want that, and I had to take care of my brother and my sister."

As Jordan tells it, he really grew up in elementary school, where he found a few key role models who saved him. He had a mentor at Inner Hope, an organization that works with youth on the Downtown Eastside, who played hockey with him, took him to Mexico, and taught him construction skills. He was involved with local community centers, was on the dragon boat racing team, learned to snowboard, and volunteered with an immigrant program. Jordan told me that he's not religious but that his mentors were often faith-based and made him "think about God and rot. Each food crop has specific temperature and humidity needs that can extend and enhance its life and quality.

Storage crops require curing at specific temperatures; more perishable non-storage crops want to be harvested in the cool part of the day, then washed and delivered immediately. We choose boxes that match the needs of the product—a shallow flat for packing single-layer, highly perishable products like ripe heirloom tomatoes, which bruise when you look at them the wrong way. Some items are field-harvested and then sorted and repacked in the processing area. Others are too prone to bruising and must be placed in their final resting box in the field with as little handling as possible.

Decisions about optimal harvest stage, choice of packing boxes, and storage temperatures and humidity are informed by whether you are direct-marketing to

the public or to restaurants, or selling wholesale to distributors or stores.

Production methods have an effect on how well a product survives transport from field to plate. Stress during a crop's growth period, the time of day it is harvested, whether it's damaged during harvest, and how quickly the field heat is removed all affect how well products store and transport.

All foods are only what their cells contain at the moment of harvest, but those cells change as the product matures off the plant, as starches change to sugars and simple sugars become more complex. Some fruits improve dramatically in flavor and sugars if given time after harvest. Many vegetables only reveal their goodness when their cells are cut or smashed or when cooked. It is the abuse at the hands of a cook that exposes tissue, releases fragrances, and brings out flavor.

and what there is after life." He considers himself lucky to have had these caring mentors who offered advice and helped him through school. They taught Jordan not to let anybody get him down. They taught him about having a thick skin, about being open to difference, about sharing and giving. He was told never to quit. He put all this to use at Sole Food.

One story, in particular, typifies what I've learned from Jordan over our time together. I used a John Deere crawler tractor with a front-end loader to fill most of the boxes for the orchard at our newest farm, on Main Street and Terminal. One evening after wrapping up work, I parked the machine by the gate, locked it up, and handed the keys to Jordan, asking him to pass them on to Seann. It was a Friday, and I would not be back until the middle of the following week.

When I returned that next Wednesday, the tractor was where I had left it, but the steel gate next to where it was parked was smashed and bent like a pretzel. I asked around a bit, but no one knew anything. In my mind I made the connection between the damaged gate and the tractor sitting nearby, and I assumed that someone had decided to try out the machine in my absence.

At the end of the day, after everyone had gone home, Jordan came up to me and, clearly embarrassed, told me that after I had given him the keys and left he'd started up the machine and reversed it into the gate. I was blown away that this young man would have the humility to 'fess up like that. I thanked him, acknowledged his honesty and the courage it took to tell me. We lifted the gate off its hinges, laid it flat in the middle of the street, and I ran over it with the tractor until it was flat again.

Jordan became our delivery person, driving a large step van to drop off orders at restaurants. If Kelsey is the face of Sole Food to the chefs, for a time Jordan was that for the establishments we delivered to. He made Alain proud.

Now, in hindsight he was probably not the best person to have out on the roads in a big truck. We might have taken a lesson from the tractor incident. He once backed into a Lexus, ran into the cheese stand that sets up next to us at the farmers market in Mount Pleasant, and backed into the toolshed being used by city workers repairing the Granville Bridge. And while our insurance rates have surely gone up, I consider it one of those costs that we need to accept if we are to give our workers a chance to become responsible.

Jordan left Sole Food in order to go back to school. He now works at the Edgewater Casino as a blackjack dealer only a few hundred feet from our False Creek farm where he used to grow vegetables. But he's taken lessons home with him. He told me recently, "Watching the life cycle of each plant has given me the knowledge to have my own garden. I'm now using up every bit of space I have at home. I'm probably planting a little closer than I should. I eat way more greens, and I always share with people. I bring home food and give some to my neighbors and tell them what to do with it."

Jordan and his father have taught me so much—about generosity, about taking care of one another, about forgiveness. Their lives are proof that, just like the plants that we grow, humans are resilient and will thrive when given proper nourishment, a sense of community, some respect, and something meaningful to do.

CHAPTER 6

Farmily

arly morning, a hint of light, ropes untied and scattered spaghettied around the truck on the pavement, tables set up, boxes unloaded and stacked like with like. We heap bunched carrots and beets until they loom overhead and then alternate with summer turnips and radishes; obscenely large heads of romaine separate onions from potatoes, while large piles of every size, shape, and color of tomato are put in place alongside basil and garlic. Greens and beans and cukes and squash fill gaps and spaces; peppers and rainbow cherry toms go front and center, berries directly behind. It's a rhythm of movement and color that takes place several times a week and that has been repeating itself for thousands of years as farmers and eaters converge at the crossroads for the great exchange.

The stage is set, produce like players washed and preened and meticulously placed. When the curtain rises we are ready, samples in hand, to entice and explain, fill bags, calculate totals, and share stories.

For urban dwellers who gather in the markets, we are the connection to the land, the bridge joining soil, seeds, weather, botany, biology, and humanity. We can talk about why the carrots are smaller this week or the strawberries so sweet.

"It's a new crop," I explain to a woman who is agonizing over a bunch of carrots. "The small ones are my favorite, sweet and crunchy, no late-stage textural challenges with these babies. Oh, and don't forget to taste one of these berries!" I add, handing her one. "It was a warm dry week, and the fruit is pure perfection." In those brief moments we exchange critical life information like how to prepare beet burgers or thin-shaved fennel salad, or what the true color of a carrot is. (It is not orange.)

"Those artichokes are so fresh they're still moving," I tell shoppers as they rummage through the stack, always pulling ones from the very bottom. Why do they do that? I always think of the bottom as the underworld, like "rock bottom," beneath, down, lower.

We few, who grow for so many, perform impossible tasks as we attempt to nurture soil, plant, cultivate, harvest, and deliver foods whose expiration date arrives the moment we cut or sever, dig or pull from the earth.

This morning, a middle-aged Frenchwoman, having discovered the stacks of Charentais melons, has been transported to her homeland. She grabs my arm to get my attention, and together we smell every melon in every crate. Despite my claims that they are all good, she insists that I choose the perfect one for her. I remind her of our "melon back guarantee"; everything we sell is testable and returnable.

I pull my knife from my belt and start slicing. Crowds gather, there are oohs and ahs, and in the midst of total chaos we all share a moment of melon bliss. As the lines build our other market staff and I entertain with other samples: a slice of Hakurei turnip, sweet peppers. (They always ask if the pepper sample is hot. As if I would *ever* do such a thing.)

Over the hours, the piles shrink and lose definition as people frantically pull and grab and bag, all while holding their place in line. This space is tiny and I struggle to move boxes from under tarps and tables to replenish and rebuild displays. I watch with amusement as some folks return items already selected in order to take from "new" stock, apparently thinking a piece of produce is fresher because it is still in a box.

One staff member's main task is to stand in the aisle in front of our stand handing out strawberries, or cherry tomatoes, or chunks of melon. Most customers politely accept but continue walking, five, six, seven, eight, nine, ten steps until the mouth has communicated a flavor explosion to the brain. They U-turn and line up for more.

In midsummer after the garlic has been harvested, we smash cloves on the hot pavement in front of our stand and watch as passersby helplessly follow the waft, seeking out every member of that Mediterranean guild: tomatoes and basil and garlic.

Sometimes, in a performance of spontaneous aromatherapy, I walk up and down the line in our stand or step out into the crowds passing a bunch of fresh mint or basil or lemon thyme under noses. A few hesitate, but for those open-nosed individuals there is a rapid and unexpected rush that sends a potent message to the brain.

"How can you afford to give away so much food?" other growers or regular customers constantly ask me.

"One small seed planted in fertile ground multiplies a hundredfold," I remind them.

It always confounds me when a new market vendor throws boxes filled with products—food they worked hard for months to produce—onto the display table and then sits down passively behind it.

There is no place to sit down at our market stand. We like to stay on our feet and meet eye-to-eye with eager eaters. Our market display requires hours to set up, and in an ongoing statement of hope and possibility, we always bring a little more than we think we can sell. We've noticed that the food does not sell if it's left in the fields or in boxes in the cooler at the farm.

Three of us work in our crowded tiny stand, but even with all hands on deck it is a challenge to sell a full truckload in the four or five hours the market is open.

"Kale of many names—lacinato, black cabbage, Toscano, dinosaur—rainbow chard, mustard greens. Collards: $3.50 a bunch, three bunches for $9!" I yell out at frequent intervals.

I remind those filling the market stand that they can mix and match, two kale and a chard, two chard and a kale, throw in a mustard green or

a collard. There are endless combinations, too many to say in a short refrain. Those customers with baskets filled only with sweet fruit get publicly admonished—by me—for not buying greens or other vegetables. Guilt trips make for an odd sales technique, but this is all done playfully, taken with humor, and sometimes even results in another bunch of kale or parsley or head of lettuce finding a new home.

The food is piled high and in your face, but folks still need a little audible encouragement to engage with the overwhelming abundance that surrounds them.

"Salad mix and loose spinach, $5 a bag, three bags for $12!"

These market mantras go on and on throughout the day. We dance the dance of farm and food, relishing the theatrical, the visual, the sensual, until piles dwindle to near nothing. Eventually there comes that point of no return, when trying to sell the last few baskets of strawberries or bunch of beets makes no sense. The day must end. And we begin the morning routine in reverse. Pack it up. Go home.

Nothing is ever entirely as it appears, and so it is with these market gatherings. Are we just trading food for money, or is there something deeper going on here?

As I see it, our food is the medium, and the message is nourishment in its most elemental and spiritual form. People ostensibly come for the food, but they really come for the connection. It's why we're here, at least. It is never our goal to serve customers perfunctorily or quickly whittle down the line. Market is our time away from the quiet, the patient, the repetitive earthbound farm tasks. This is our time to exchange energy. We want our stall to be bursting and spilling over with food and people, to have a constant buzz, to be humming with vitality. The jiving and joking, the stories from the land, the constant engagement—this, like dining in the restaurants we supply, completes the circle. The sampling lets us eat with our community.

We sell at three Vancouver markets and for a time operated a retail outlet at Granville Island that was open seven days a week.

Our policy at Sole Food is to include a staff member from the Downtown Eastside front and center at every market. Some staff immediately thrived in that role, and some have grown into it. Still others continue to struggle with putting themselves forward in public, with having to engage, with developing self-confidence, and with accepting public perception. Customers, too, sometimes seem uncomfortable with having to deal with someone who does not fit the normal farmer profile or someone whose public presentation may reflect socioeconomic difference or a life of hardship.

Allowing the public the opportunity to directly engage with people they might not normally engage with is a powerful service for both our staff and the community. We all get to look at our fears, our judgments, and our stereotypes when someone out of our circle offers a sample of a tomato or strawberry or pepper. Is it safe? Who are these people? What is Sole Food? This small, momentary public exchange and the questions it engenders offer us all an opportunity to move toward understanding and acceptance. Once again, we see fresh food as the medium, the messenger, the means by which all of us, no matter where we come from, come together.

———

Our work in the markets, and supplying them, has, over the years, highlighted in unique ways the struggle we face in balancing our social agenda with our agricultural one. It's not only the public who asks, *What is Sole Food?* In many ways, our two primary goals—to provide meaningful employment to individuals who have challenges and to create a credible model of production urban agriculture—rub up against each other and can be contradictory. While we try to harmonize the two conflicting goals, I suspect that this conflict is a reality that will always be there. Still, we work hard on the problem. And we've got a hard worker who spends a great deal of time trying to figure it out.

We hired Lissa Goldstein in 2012 when we were expanding the project, as site manager for the farm at Pacific and Carrall. Seann and I conducted a job interview with her via Skype, both peering into his iPhone while standing at our Astoria farm as she smiled out at us from the East Coast of the United States. I could barely hear above the din of traffic or see through Seann's shattered phone screen, but what came through was that this Dartmouth-educated young woman was strong, supersmart, and solid in her intent and had a few years of both farming and community development experience.

Perhaps more than anyone else at Sole Food, Lissa has come to know there is a fine line between providing real opportunities for people to learn new skills and get healthy and pandering to a sense of entitlement that exists among some people in this community where social service agencies are so abundant. How much can we expect from our staff with the challenges they have? How much is realistic for them to expect from us and from the organization? In her role as director of operations, Lissa feels these contradictions acutely. She now holds the responsibility of the day-to-day management of our staff and farms. And it's there—in

the rows, under pressure, facing the elements—where the challenges most often appear. The devil is always in the dirty details.

"Most places, all employees are treated alike, but it does not work that way in this job," Seann reminds me. "Some guy is suicidal and another person is having anxiety issues, one is doing heroin, and another doing crack. Their personal stories play out so much in the workplace."

After a week of rain at the end of June 2013, the temperature rose to the mid-eighties and stayed that way. Our radishes took off, with six or seven beds maturing weekly instead of the usual three. We were selling 350 bunches of radishes every weekend, which meant we were picking a lot of radishes. On the parking lot there was no respite from the heat. There was no cool house to retreat to for a power nap, only the urgency of getting those radishes out of the beds and into a cooler. And Lissa felt the heat too, of course—plus more. She tells me, "I always feel like my anxiety skyrockets in these situations because we have staff for a limited number of hours in the day and four hundred radishes to pick before it gets too hot."

One morning in the midst of this rush, when by 7 AM the sun was already oppressively hot and Lissa was already feeling the stress, Donna approached her to ask about holiday pay, of all things. The timing seemed off, and Lissa responded, "Yes, you're going to get it, but we don't actually have to pay for holidays." The information was more than Donna needed—"Yes," would have sufficed, Lissa admits—and Donna grew angry. The discussion quickly deteriorated into an argument and then tears.

It seemed that something had been building in Donna. And as the argument settled, she pointed out that since she'd first been hired, many of the employment-support pieces of Sole Food had disappeared—the staff meetings and farm walks, especially. To her, the focus seemed to be exclusively on production. For Lissa, this was a really difficult situation. As she tells me, "I was forced to deal with my vulnerability, the stress of a large staff, and the difficulty of balancing commercial farming with real supportive employment. I was also forced to deal with the real question of what Donna was saying. What had gone wrong? How could we balance these two things better?"

This collision forced us to look at what was missing. In late summer 2013 we initiated more regular staff meetings and check-ins. Around the same time, an employee representative was elected by the staff and started attending manager meetings and reporting back to the staff. These seem like small changes, but they were, in the end, profound. They allowed everyone to participate more fully in how things were run.

That fall, as production began to slow, we organized a day for all of our staff devoted to sharing ideas and concerns, hopes and challenges for the future of Sole Food, and team building. Donna apologized to Lissa and there was a palpable sense of healing and relief for both of them and for the whole group.

"Having staff meetings and being able to talk about their needs has been great," Seann later noted when reflecting on those earlier times. "They are now way more respectful about calling in when they can't make it to work. We haven't fully hit it yet, but including them and allowing for a staff feedback loop, and not singling anyone out, makes them feel like they are a part of the day-to-day decisions and process of how the farm and the business runs. We are doing a good job, but there are a few more steps to go."

Seann and Lissa have been doing regular performance evaluations with staff and filing them in sequence so that everyone can check in on their progress. Our staff members now have the opportunity to create their own job descriptions. And they'll tell strangers in the middle of the city that they're farmers. Before, there was a lot of hand-holding; now I see real independence and initiative on the farms and a sense of responsibility. Part of this new way of operating has involved holding us, the managers of Sole Food, responsible, too.

One day, for instance, Alain reminded us that many on the crew were coming to work with empty stomachs. "I bet you'd get a lot more work done if they had something to eat in the morning," he suggested. And so it happened. The company Nature's Path donated cereal, and the staff made a list of other food items they wanted to have in the staff area. Now there is milk and cereal and, thanks to another donation, Salt Spring Coffee. We have an arrangement with a local bakery that brings us fresh bread, so staff can take bread home, and we have a source of fresh eggs.

As a business, Sole Food Street Farms has grown, and today we are respected in the community. Our products are not the cheapest, but the quality is excellent, and our staff carries a level of pride in what they have accomplished and who we are in the neighborhood. We have never had a goal of trying to change anyone; we just wanted to create a stable enterprise that can provide meaningful work year after year. Unlike many social service agencies, we don't train people and move them on. We are proud to have several people who started with us seven years ago still involved. Just as in any work environment, to be successful everyone needs to feel fulfilled and excited to be at work. The best fertilizer is the farmer's footsteps on the field. When the farmers are happy the food tastes better.

Happiness is not a constant, of course, and we must admit where we've fallen short. Not everyone always feels fulfilled. While much energy has gone into supporting the staff who come from the neighborhood, those of us who are not Downtown Eastside residents—Seann, Kelsey, Lissa, our office manager Tabitha Mcloughlin, me—sometimes we fall through the cracks. The founders and the core management team are holding a lot together. Often our work is behind the scenes, less visible, and burnout can creep in unexpectedly.

It is fairly common for founders of an organization to find themselves filling roles that no longer allow them to do the very work that inspired them to begin with. Seann, for instance, does all kinds of tasks that are essential but invisible, and I know he has experienced, as have I, a kind of disconnection from the daily work with the crew. Lissa is on the farms every day guiding the crew and the agricultural operations, and, as she experienced with Donna, that comes with its own complicated pressures and disappointments and sometimes tears. Tabitha, whom we brought on in 2013 to do bookkeeping, communications, and events, and to look after the millions of details that we could no longer keep up with, is in the office doing administrative work. "We are so focused on

Lissa

creating this environment," Seann says, "but as you move up and you are looking for feedback and you can't get it or you don't have a peer group, it's hard for all of us."

As with almost every aspect of life and community, solving problems comes down to communication. When the five of us get together to talk and we can discuss our personal challenges, we tend to feel better. We remind each other not to separate ourselves from those we are trying to support. And yet sometimes, our responsibilities do separate us, and speaking from personal experience, this can wear a person down.

Seann and I have considered Sole Food becoming a worker-owned business, which would take a lot of the pressure and expectation off us. "But," he admits, "it's a double-edged sword. Our staff wants more responsibility, they want to be held accountable, but sometimes when you do hold them accountable they bristle. We are supposed to be here for each other." Seann is not alone in expressing these occasional frustrations. But just as soon as we identify a problem, we're reaching for solutions. As Seann says, "This project has been successful because it has been built on the backs of Downtown Eastside residents as the core and reason for the project. As soon as you take each other or any of the work for granted, it falls apart."

Sole Food's commitment to open communication, about the good and the bad, our trials and errors, is the best reminder that it takes all of us—founders, managers, staff, and the whole community—to make this work. I won't lie; acknowledging our weak points, having them revealed to us by our co-workers, has been difficult, but personal reflection and learning how to live with one another is what we are here to do.

We talk about balance on our farms and at our markets, but what does it really mean? I've learned in my life as a farmer, an employee, and an employer that perfection cannot be the goal, and as such we will always be asking ourselves difficult questions. We will never have all of the answers. We must accept that we can never get comfortable. The human aspect of our work is just like the agricultural aspect of our work; it is always changing and evolving, and it will always require a beginner's mind—a willingness to see things anew—to work with it.

———

With her matted blue-streaked hair, pierced nose, tattoos, and patched blue jean jacket, Nova might be mistaken for just another one of the lost souls walking past the farm on Hastings Street. Look a little closer and you'll see there is soil under her fingernails, and the telltale red- or green-stained hands of someone who's been harvesting strawberries or tying tomatoes. She tells me her name grew from Nova to Novocain during a time she was doing peer support and emergency harm reduction. "They'd be having a bad day and I'd take them out, they'd rant and then feel like they had a dose of Novocain on life."

In 2003, at sixteen years old, Nova fled a group home in Alberta and traveled to Vancouver, hitchhiking and taking the bus. When she arrived she made her home on the streets and her solace in crystal meth. Nova refers to herself now as a "retired Granville Street kid" who supported herself and her habit through panhandling. "I developed a street family tribe," she's told me, "and we stood by each other and eventually helped each other recover."

In 2012 Nova came to work with us at Sole Food. She refers to her Sole Food tribe as her "Farmily."

I've searched for language that describes the work we try to do and too often I fall short. "Farmily," the product of a beginner's mind, says it

Michael

Nova

all. Many of us spend our lives searching for our tribe, trying to re-create what we had or did not have in our primary families. Some find that connection through marriage and kids, others by joining gangs, some through clubs or schools or churches. The native peoples who inhabited these lands long before us found their connection through the earth; they worshiped it and were educated by it. They didn't require schools or churches. Their whole world was both.

"Farmily" implies community in all of its manifestations: soils that contain diverse communities of microorganisms, dynamic communities of plants and animals, the farm as part of a broader ecology, a vibrant community of farmers providing husbandry and stewardship, and the wider human community that depends on the farm as a gathering place and for its basic nourishment.

Nova tells me that she stiffens when someone refers to one of the farms as a "garden" or our staff as "gardeners." It's a sentiment I would never have expected when we first started this work, but it totally resonates with me. If you've been out there in the middle of acres of boxes full of food, digging beds or transplanting, or it's a market day and there is a truckload to be harvested, you definitely don't think of yourself as a gardener. Through workers like Nova, coming to this with fresh eyes,

FARMILY

I've had my own sense of who I am clarified and reinforced—it is deeply satisfying to recognize yourself as a farmer, to carry that distinct identity, to think of your work of planting seeds and producing lots of food as a profession, not a hobby, and to know that your hands are growing for a hundred times your own self.

On Thursday afternoons we sell at the Yaletown farmers market. Our stand is sandwiched between the cheese people and an artichoke stand. Nova stands beneath the classic street sign we place out in front of our stand that says SOLE FOOD ST. FARM. She hands out samples of cherry tomatoes or strawberries to passing shoppers. And I find it a remarkable sight: This young woman who used to spend her days on the street panhandling, whose anxiety around people was palpable, now smiles and carries on comfortable and confident conversation with total strangers.

"All of us at Sole Food have come from such heavy roads, but we all survived," Nova tells me. She's now clean. She says she's very strong into her sobriety. She's a mom. And she finds herself feeling more and more grounded as the years go on. "It's strange," she says, "that I've been working in one of the biggest cities and yet I'm involved in a country profession. I was a floating spirit, and now I've found my passion and where I want to go. I've learned about plants and how food can be medicine, the real medicine, and I want to share that with my peers."

———

Once a week we fill cotton bags with an array of farm products that might include salad greens, spinach, peppers, tomatoes, parsley, carrots, beets, chard, melons, radishes, ground-cherries, and cipollini onions, pack them into blue plastic bins, and load them into the truck for the ten-minute five-city-block trip to the corporate offices of Vancity, Canada's largest, and I imagine, most forward-thinking community-based credit union.

We drive, Seann behind the wheel, into the underground parking area, pull up near the elevators, remove the load, and take the elevator one story up to the security desk, where we all receive our visitors badges.

Signed in and secure, Kenny and Seann and I carry the bins up to the second-floor lunchroom where these food shares are distributed to Vancity staff. As with so much at Sole Food, we make for an incongruent scene: dirty pants and muck boots on immaculate green and purple carpet welcomed by well-dressed employees of one of the most successful financial institutions in the region. We're delivering boxes and cloth bags with food that is only a few hours out of the fields. From the window of the lunchroom, you can look down Terminal Avenue to Sole Food's orchard at the corner of Main, and beyond to the roads, railways, and buildings that stretch eastward into Burnaby, Coquitlam, and beyond into the Fraser Valley.

Community supported agriculture, or CSA, came to North America from Europe in the early 1980s. Originally CSA was a form of social agriculture whereby community members threw in their lot with the farmer by purchasing a share of the farm's annual budget; members would then receive a share of the farm's production throughout the season. At its heart, the CSA concept had the goal of using food and farming to provide new models around property ownership, farm management, economy, and food security.

In North America the CSA model, in its purest form, attempted to address the problem of one farmer investing his or her own funds and taking all the risk in order to grow food for an entire community. It's a noble idea that everyone who eats should share in the risks of how food comes to them, but over the years the CSA model has become less social and more like a simple subscription service—just another way of selling and distributing food.

In the early 1980s I started one of the earliest CSAs on the West Coast. I too had lofty goals for our program. I planned to close our on-farm retail store, stop attending farmers markets, and funnel all of our food into a membership CSA program. I was desperately trying to find ways to integrate the farm with the new suburban community that had moved in around us. But the community was not ready for my agrarian and social ideas.

Life was moving pretty fast for folks in that neighborhood, and in the end I concluded that busy people didn't want to go to potlucks or farm tours or volunteer to distribute food shares. They wanted convenience, and they wanted low prices. We continued that program along with all of our other more traditional sales outlets. While the CSA did provide a dependable advance source of income, reduced some economic uncertainty, and forged relationships with some community members, it was for most people just another way to trade money for good food. Some members felt that getting to know the farmer, connecting with the land where the food was produced, and participating in the process in small ways were important, but they usually could not make the time to participate in what might have offered a hands-on experience.

Now, twenty-odd years later and thirteen hundred miles north, we are doing our own Sole Food version of the CSA concept. And while I feel strongly that people should buy products because they are the best products available, rather than out of some sense of charity, I know that charity happens. I have never liked the idea that anyone would support a farm because it is organic or biodynamic or because it employs people who need work or because it reflects some mission or belief system. If the food is not the best food, then find someone who grows it better. No farm should be able to survive simply because it has a good story. Even if that story is as good as the one we're trying to tell.

I know that Vancity loves both our story and the excellent quality of our food. You can see Kenny's photograph on the company's advertisements plastered on the sides of city buses or next to the teller in one of their twenty Vancouver branches or at the ATM machine. Jordan is in their print ads, and during a commercial break from a hockey game on television my son Benjamin saw footage of one of our farms in a Vancity advertisement. Vancity has been incredibly supportive of our work in so many ways, so delivering directly to their headquarters completes the circle, allows us to personally deliver a piece of ourselves, and brings the relationship down to the core element, good food.

And while I so appreciate hearing people tell us, "We love what you're doing for folks on the Downtown Eastside," as a farmer, the greatest acknowledgment I can receive at the market, or from a chef, or during CSA distributions is, "We love your carrots and greens and melons."

We sell and distribute beautiful food at markets and through our CSA. And many people do seem to love the produce that we grow. But what's not always obvious to those who purchase and eat our food is a lesson we learn each day, each season, year after year on the farm. And if someone—a customer, a visitor to the farms, a Vancity employee—gives me the time, it's a story I like to tell. Agriculture is the confluence of seven thousand years of trial and error. And we've known our share of error.

This seven-thousand-year agricultural experiment has mostly taken place on open land, in fields and pastures. Agriculture did not evolve on pavement, between buildings, in the remains of demolished factories or homes, or in soil that has become contaminated; nor did it evolve to be carried out by the hands of those whose lives, like the cities they inhabit, are in a continual state of being torn down and rebuilt and recovered. Growing food on pavement in cities represents a new type of agricultural experiment, one that does not have the benefit of those who came before. It is fraught with false starts,

Giving Versus Selling

It has never been our intention to give food away. When Seann and I first began meeting to discuss the Sole Food project, the ideas and goals on the table included growing the food, feeding the neighborhood, providing jobs, teaching people to farm, and more. We knew we could not achieve all those goals, and it became clear that the most important one was providing meaningful work. To provide jobs meant a monthly payroll that could only be met by selling food, not giving it away.

In any farming operation there are products that you cannot sell—market returns, cosmetic imperfections, items that are bruised or damaged. Our goal is to keep these inferior products to a minimum. Whenever possible, we process "seconds" to add value to them, or we try to sell them at a discount in order to cover the loss.

In an unconventional form of observational accounting, I evaluate my farm's financial well-being based on how much product I see sitting on the compost at the end of the day. I view the compost pile as a soil savings account that accrues value with time, but I also see it as the place where I can visually register my losses. It isn't that the compost is not valued in and of itself, or that products going into it do not add value to our soils; it's just that there is so much embodied energy in the long and tedious farming cycle, that those products decaying on the pile always represent some financial loss.

Due to municipal restrictions and the scale we operate on, Sole Food is prohibited from composting on our farm sites. As such it has been difficult to recycle old or damaged products, weeds, and completed plantings back into our soils.

monumental mistakes, and new discoveries. We at Sole Food know this as well as anyone.

As well as it went, the first season at the Astoria was chaos. Most people on our team were struggling with their addictions and other personal challenges, which added immensely to the challenges of a first season of farming.

The first day of planting I arrived in a Volkswagen van filled with transplants we had grown at Foxglove Farm. I'd brought rainbow chard; lacinato, red Russian, and Winterbor kale; red leaf and romaine and Lollo Rosso lettuce. I had seed too for French Breakfast and Cherriette radish, arugula, salad mix, and spinach. It was a simple crop mix. I knew

FARMILY

This has presented a severe limitation in our ability to complete natural biocycles. Adding insult, we often have to pay to have valuable organic materials hauled away, in some cases to places where those materials will not be composted.

When at all possible our seconds and market returns go to local food banks and soup kitchens. It may sound strange and contradictory, but while I am thrilled to be providing food to those in need I am equally concerned to discover how much of our food we give away each year. Our financial security and ability to keep operating are dependent on selling as much of the food we grow as possible and at the highest return. So while there is an ethical imperative to give food away, the financial realities of our farms are ever present and demanding.

This becomes more acute on a smaller farm. The larger the scale of production, the more value is realized by higher volume; the smaller the scale,

the more essential it is that every bunch of carrots or beets, every pound of greens, every radish and tomato realize its greatest value.

It is one of the great contradictions of what we do. We know that the fresh food is desperately needed by the very population we employ and all those who are living at poverty levels. We know we need to sell all the food to pay the bills, and we have to accept the incredible challenges inherent in attempting to marry a biological system with a market economy.

This last challenge is significant as we try to support the biology and fertility of soil and the well-being of our staff and the community while working within a market economy that does not always place value on soil organisms, mental health, and social inclusion. How do we go beyond the traditional bottom line and reconcile these two divergent pulls?

that the crew would not have the skills to grow a wide range of crops, and we had not yet developed our markets or the CSA, and so I focused on those crops that would have a high probability of success. Our staff and this community were used to failure in other aspects of their lives, and for the project to sustain itself beyond that first season we needed immediate, solid, visible, and edible positive results.

And while it was a real shock for those young plants being pulled out of the warmth and protection of our rural farm's propagation house and thrown into a cold spring day in a foreign urban world with sirens blaring and people staring, it was far more of a shock for me. I looked around the parking lot with wood boxes filled with soil and a motley crew of

— 133 —

down-and-outers all standing in the rain, water up to their ankles, and I wondered what I'd gotten myself into.

But still, we gathered around and I gave a little speech, like the coach before the big game—only this team had never played this sport, and they looked at me and the plants and the planting beds with varying degrees of confusion.

We marked the planting rows, had people drop plants at the pre-scribed spacing, and started planting. Sure that I had explained things carefully, I put my head down and started putting plants in the ground. Completing a fifty-foot bed of chard, I looked back to find the whole crew still clustered together at the start of the rows, contemplating and plant-ing each plant as if it was the only one they would have to plant that day.

It is inconceivable to me that only six years later some of those same individuals, men and women who had not previously held a job for any significant length of time, have developed their skills so much that Sole Food was voted by *Vancouver Magazine* as the top food pro-ducer in the region.

FARMILY

Success is a relative term in the world we work in, and so I use it with care. Awards and acknowledgments have come, but we continue to struggle with the day-to-day details of running a social enterprise. We face the ups and downs of our staff combined with the unpredictability of a biological system that is constantly changing and influenced by natural—and not-so-natural—forces beyond our control.

CHAPTER 7

Rewilding

nowmobiles, eighteen-wheelers, an elephant, buses, cars, and rickshaws are stalled in a mass international traffic jam. Drivers, passengers with luggage in tow, a cacophony of multilingual verbal insults flying, radio music blaring, and horns honking add to the chaos and confusion.

The mock traffic jam, created for the 1986 World Expo—which was themed "Transportation and Communication: World in Motion, World in Touch"—took place where our carrots and beets and peppers and greens now grow.

The year 1986 was a big one for Vancouver. Prior to the World Expo it was a backwater town, spectacular in its dramatic mountain and sea context but little known to the rest of the world. The billion-dollar Expo, however, brought twenty-two million people to Vancouver and initiated the conversion of a vast tract of industrial wasteland that existed along False Creek. When the Expo was all over, Hong Kong tycoon Li Ka-shing, through his company Concord Pacific, bought the land for $320 million. Some would say this was a paltry sum, knowing the tract's current value. I've heard comparisons between the sale of this part of Vancouver and the purchase of Manhattan from native people for beads and trinkets. It's a fine comparison until you consider the fact that to develop the heavily contaminated False Creek land would require a staggering amount of money for remediation.

Since the sale to Concord Pacific, much has germinated and grown up and around the acres of pavement that were once home to the Expo, including high-rise luxury condominiums, Science World, the Vancouver Canucks' Rogers Arena, and the seawall used by bicyclists, runners, and walkers. What remains undeveloped sits in politically charged limbo, awaiting its final crop of concrete, glass, and steel.

STREET FARM

Here's where we saw an opportunity. The former Canadian Pacific rail yards left behind contaminated land that was eventually capped with asphalt and turned into a series of huge parking lots. And in spring 2012, one of those lots, known as Sub-area 6C South, saw the birth of our largest farm.

Despite the ease involved in securing the site through our dealings with Terry Hui, building the farm infrastructure there was a monumental effort—twenty-six hundred used shipping pallets had to be found and brought on-site, covered with landscape cloth, and carefully lined out over the two-acre space allotted to us. The pallets were outfitted with our new SOLE FOOD–imprinted plywood collars, open at the top and bottom, which created a box. Soil, made from organic waste from the region, was then pumped into the boxes. By late June we had seeded or transplanted a substantial portion of the lot.

There was a lot of hoopla around the establishment of this farm. The sheer scale of it was dramatic. Two acres is the largest site we'd farmed in the city. The site is also our most highly visible, located directly next to many of Vancouver's most visited attractions and venues. You can see

it from passing commuter trains, and the farm is flanked by those bike and walking paths at the seawall.

We share the five-acre parking lot with the annual Underwear Affair (a ten-kilometer run that raises money for cancer); staging for the Sun Run and Vancouver Marathon; a street hockey event; dragon boat storage; parking for people attending sporting events, concerts, Cirque du Soleil; film crews; broadcasters for the FIFA Women's World Cup; police motorcycle drills; and the luxury coaches and trucks that transport equipment and performers such as Justin Bieber and Madonna. This may be one of the most visible farms in the world.

And its conspicuousness, so often our goal with urban farming, almost immediately threatened to sink the whole enterprise.

Soon after the site was established, with seeded crops just starting to emerge, transplants putting on new leaves, the city of Vancouver environmental services vehicles arrived to test the water in and around False Creek. The results of those tests showed fecal coliform rates that were off the charts, in the tens of thousands, with normal rates being in the forties.

Concerned by these numbers and wanting to establish the source of this pollution, the environmental service officer assigned to the case made the immediate assumption that the contamination must be from the farm. We were the new kids on the block, involved in an activity that was unheard of on this scale in the city, and as such we must be creating the huge spike. Excited by "discovering" this obvious connection, the officer pointed the finger at the soil in our growing boxes.

One of our major funders got wind of this and panicked at the possibility of negative media coverage, and suddenly Sole Food, its expansion, and its biological integrity were placed literally under the microscope.

We needed to calm the emotions of our funders, overanxious city officials, board members, even potentially the loan officer at the bank considering our loan request. With this in mind, when the report came out I immediately sent off an email expressing that it was premature to jump to any conclusions, especially with as little information as we had.

My instincts told me that you would have had to tap directly into actual fecal matter to have gotten the kind of off-the-charts readings the city was claiming, and none of the plants we were growing could survive, let alone germinate, under those levels of fecal matter. To head off a potential media frenzy, we voluntarily submitted soil samples to a lab in nearby Burnaby to be tested for fecal coliform rates. I explained this step in the memo, as well.

This moment at False Creek illustrated a classic collision between the urban environment we were working in and the activity of

production farming, which normally takes place far from most cities and from urban consciousness. The only agricultural operations that could have produced those high fecal rates would have been concentrated livestock operations. The only animals on the site were birds and the occasional neighborhood dog.

Eventually we received the results of the analysis done on our growing soil. The fecal coliform and *E. coli* levels in the lab test results of our soil were at the lowest test parameters that could be measured, 20 MPN/100 grams.

This proved that there was no relationship between our soils and the spike in fecal coliform rates that the city came up with, but it did not answer the question of why False Creek experienced such a dramatic spike.

No one—well, I did, but no one else—seemed to notice the multi-million-dollar yachts anchored offshore from the farm in False Creek disgorging raw sewage directly into the water. Likewise, no one seemed to want to acknowledge that Vancouver still has numerous underwater sewage outfalls that at times send sewage and toxic storm water into surrounding waterways, or that the capital city of Victoria, tourist mecca, seat of government, spews hundreds of thousands of gallons a day of sewage directly into Georgia Strait.

After the uproar settled down we went back to the work of growing food on the shores of False Creek. We voluntarily placed straw bales end-to-end along the entire creek side of the farm to catch soil and slow any drainage coming off the farm. In 2013 students and faculty from the British Columbia Institute of Technology installed a fifteen-hundred-square-foot constructed wetland on pavement at the bottom of the farm to funnel all of our runoff into a beautiful planting of sedges, rushes, and fescues, which capture sediment and naturally absorb nutrients.

To the uninformed eye this beautiful little piece of constructed wildness might look like a bunch of unkempt weeds, but what emerges from it and into the storm drain below is crystal-clear. The results of this little piece of biomimicry represent but a tiny drop flowing into a creek we call False, but each drop leads into a vast sea. We do our part.

————————

In the summer of 2015, three years into our lease with Concord Pacific, the farm was thriving with crops growing in new plastic boxes in place of the old delaminated wooden ones. We'd set up an office and packing area and staff hangout zone, and there was a general feeling of permanence on our vast expanse of asphalt. Although conditions were hot and dry that summer, the staff was well organized and highly productive. By fall all of us had the feeling that often comes in the fifth or sixth year of a new farm enterprise: Systems are in place, things are settling down, and some semblance of stability finally arrives.

Yet still, something was worrying me.

Early in 2014 Concord Pacific had turned over the management of the parking lot to a company called WestPark. Parking is big business, and the man at the helm of WestPark, John Laires, may have looked at the thousands of boxes of food crops on that perfectly paved lot that was taking up 375 spaces for parking cars and cringed at the dollar value lost. Sole Food's direct financial contribution to Concord Pacific was nothing—again, just a dollar per year—and I'm not sure John knew or considered that our presence represents a property tax reduction as well as an immeasurable contribution in positive community and public relations.

After the transition to WestPark, there was some confusion about whom we now answered to. Before this, we discussed any questions or issues directly with our landlord, Concord Pacific, but after WestPark took over—and unbeknownst to us—John had taken over the role of our liaison. John was enthusiastic about his role as CEO of a relative newcomer to the Vancouver parking scene, and on occasion his approach ruffled the feathers of members of the Sole Food staff.

Growing food and nurturing people's lives require different instincts from parking cars, and it became clear in subtle and not-so-subtle ways

The Parking Space

An average parking space is 9 by 18 feet, or 162 square feet. If you are working with biologically active soils and you are reasonably skilled as a farmer, a 162-square-foot parking space can produce 450 to 500 pounds of food in a four- to six-month growing season. Five hundred pounds of food at an average retail market value of $3 per pound is worth $1,500. There are an average of 150 parking spaces in an acre; there are approximately 9 acres of underdeveloped land around our False Creek site, totaling 1,350 parking spaces. Multiply 500 pounds times those 1,350 parking spaces and you get 675,000 pounds of food in a four- to six-month season with an average retail market value of over $2 million. Of course, all of this is based on ideal conditions, a high skill level, and a guaranteed market for products produced. But there's a lot of potential in a parking lot.

that John may have viewed our farming work and our relative paltry financial offering to the landowner as of little value. As such, we became subservient to the parking area we shared the lot with, which is often leased for a week or two by movie companies or sports events. WestPark could also rent out individual spaces for a few hours at a time at $20 a pop, and when those opportunities required space we intended to farm, there was some conflict.

In fall 2014 during a meeting with Concord Pacific to renew our lease, we were asked to give up land for a walkway to better serve the patrons of the WestPark lot as they made their way to and from their cars to the nearby arena and stadium. The new walkway carved off and made less accessible one section of our farm. WestPark also fenced off the entire site, which at first felt strange and isolating, but in the end actually helped us with ongoing security and theft issues. Even so, there was a general feeling that we were the inconvenient bastard child conceived during a past time when parking spaces didn't have quite so much value.

In fall 2015, after the summer of stability and great harvests, West-Park asked us to give up the now isolated triangle of our leased land in exchange for a smaller piece on the other side of the road, land that they could not use for parking. We were in the peak of an intense fall harvest season, and I rarely carry a cell phone. John Laires was growing frustrated after trying unsuccessfully to reach me, and I was quietly brooding over the potential loss of valuable growing space. By the time we made phone contact, the tension between us was obvious. John stated his concern: Hockey season was about to begin, and Canucks fans needed more space to park. Our piles of shipping pallets and soil and the boxes where we were growing hops for a local brewery were in his way.

During that conversation I challenged John on his aggressive approach and questioned how he had been treating some of our staff. I feebly, and mistakenly, attempted to defend our work, trying to convince him that Sole Food farming was as important as parking cars.

At some point I pulled back from the heated conversation. I could sense it was going nowhere. So I hit the reset button, remembering that our presence on that parking lot was by the grace of the landowner, that I was now speaking with their representative, and that the most important thing I could do was to be a good neighbor.

The conversation settled down, I accepted the outcome, and we started working out the logistics to support the change that had been decided long before we connected on the phone. We agreed to meet at the site, look over the cleanup and movement of our materials, and review a possible replacement area. On the appointed day I arrived at the Concord Pacific offices carrying a peace offering, a cloth STREET FARM bag filled with Sole Food peppers, garlic, tomatoes, greens, and a jar of our newly canned salsa.

There is a bit of poetry in how things have gone at the False Creek farm, another tale—if a little distorted this time—of birth and death and renewal and impermanence: industrial rail yards to False Creek paradise, paved and parked, then turned into a farm, then, in places, back into a parking lot.

Although my primary means of transportation while working in Vancouver is a bicycle, I do drive cars and trucks, and I realize I am as much a part of the problem as anyone else. But I also think the fossil-fuel automobile is on its last legs and that the only good parking lot in our future may be the one where you enter, park your car, and leave it there forever. Or there's always the one I'm most familiar with—a parking lot that's been converted for growing food. Right now we still have two acres of white-lined pavement at the Concord Pacific site, still our largest farm, still green and lush and abundant and productive.

We grow a lot of greens at Sole Food: butter, romaine, and leaf lettuces; spinach, arugula, parsleys, and chicories; leaves of mizuna; red flat leaf and green frilly mustard; savoyed Tatsoi; and multicolored baby and large kale and chard leaves. We produce different mixes of these greens, including a popular and visually attractive stir-fry mix. Throw it into a pan with a little toasted sesame oil and garlic, and all is well.

We pick the leaves at baby or teenage size, just when they have developed a little spiciness, but before they get too hot or lose their tender texture. The mixed colors and textures of the reds and greens and the frilliness and blistered savoyed leaves provide a beautiful visual treat even after the inevitable wilt of cooking.

Local chefs love these mixes. We do all the work of growing, harvesting, mixing, and even washing, creating a dynamic visual and delicious addition to any meal. The chefs get to revel in the adoration they receive when those mixes are put on the plate. Occasionally there's the appropriate homage to the grower who produced them when the farm's name appears on the menu. Chefs are grateful because we do our best.

Yet sometimes things go wrong.

During the winter of 2012, down to a skeleton crew, we relied on an affable young man who had been working with us as an intern to handle the harvest and delivery of our winter restaurant orders.

One day our intern dutifully harvested an order of stir-fry mix for one of Vancouver's most notable fine-dining restaurants. While cutting the greens with a serrated knife, he also cut his finger. He wrapped the cut with a Band-Aid and returned to complete the harvest. Somewhere along the way, unknown to the intern, the Band-Aid fell off his finger and integrated itself within the box of greens.

He walked the few blocks to the restaurant, delivered the order, and returned to the farm satisfied with his work. At six o'clock that evening, Seann received a call from the chef de cuisine saying that he had opened the box of greens and found a bloody Band-Aid.

Thankfully the Band-Aid was discovered before it made its way into the dining room to be served on someone's plate alongside grilled lamb chops or poached halibut.

Knowing they needed the greens for that evening's dinner service, Seann returned to the farm, harvested another box of greens, and delivered the new box to the restaurant.

We've all heard stories about food being dropped on the floor in restaurant kitchens only to be picked up and put back on the plate, and other less-than-savory tales of life in restaurant kitchens, but to be responsible for such a gross error unnerved me. I did not hear about this incident until weeks later, Seann was probably a little hesitant to tell me. The episode was an accident, there was no malicious intent, no desire to

bloody our reputation, just an honest mistake on the part of an inexperienced intern and a Band-Aid that would not stick.

Band-Aids are designed to temporarily cover, protect, and help heal. This Band-Aid did the opposite: It exposed a weakness in our harvest and post-harvest practices, forced us to reevaluate the training we were providing our staff in food safety, and reminded us how much oversight was required of our staff and interns.

———

I arrived at the farm on False Creek one day and noticed this native guy working with Alain harvesting radishes. I learned he was a new hire, so I went over and introduced myself.

I assume that everyone has had the experience of meeting someone and sensing an affinity. I liked Cam immediately; there was a groundedness and a calm about him.

Seeing Alain, who not that long ago would not have known a radish from a beet, so patiently and graciously instruct Cam on which radishes were ready and how to harvest them was inspiring. In the midst of the intensity of our enterprise—when mistakes like the Band-Aid happen—it is easy to forget why we do what we do, and whom we're doing it for. These simple scenes between our farmers are reminders, little momentary gifts.

It is awesome to witness such confidence among those who have been with us for a while, and to see their comfort in newly acquired skills. Give someone the chance to teach and he realizes how much knowledge and experience he himself has acquired. You don't know how much you know until someone asks you a question or wants to know how to do something. Recently I heard Seann describe "mentoring" other young farmers, and I remembered that it was only a few years ago that he was introduced to farming himself.

This is what I, and all mentors, should hope for, but it can be difficult to let go and allow those you have mentored to go on and surpass you. But it's what keeps the world spinning, allows life to continue. It's as if you spend your whole life getting to the front of the line, and as soon as you reach it you are ushered to the back where you began. The circle closes again.

I like the natural humility I feel from Cam, something I always admire in others. As I get older I can see that you either voluntarily incorporate humility into your life or life will do that for you. Life will force you to accept that ultimately you are not in charge. Although as I think about him, it may be that what I perceive as humility in Cam is, partly, his having been beaten down by life's circumstances.

Cam tells me that he came to the Downtown Eastside from Lac La Biche, Alberta, a small rural town 140 miles northeast of the capital city of Edmonton. Like so many First Nation rural refugees arriving in Van-couver, he had no money, and this neighborhood, now considered the largest urban First Nation "reserve" in North America, was a natural place to go for free meals, shelter, and cheap beer.

It turns out that the parking lot where our first farm is located was where Cam hung out when he arrived. "I used to go to the Astoria and buy beer and then sit in the parking lot next door to drink," he's told me. Sometimes one of the hotel's staff would come out and yell at them. Since Cam's been working for us, he's heard from an old friend, "It must be weird for you to be working here now instead of drinking."

But it's not weird. And Cam has told me that he finds the initiative behind Sole Food really impressive. I love the way he talks about the place: "There used to be beer cans, rigs, people would stop and fix there, there were broken crack pipes, and occasionally you'd find the odd toonie and feel like a millionaire when you'd

Cam

pick it up." It's amazing to think what $2 can feel like to someone who's got nothing.

Living in the neighborhood Cam supported himself by selling weed and counterfeit cigarettes. He's Cree, but he says he didn't learn anything about his culture until he was in his twenties and in prison. He learned from other inmates who were also Cree. But in prison his associations remained focused. As he's told me, "I did a home invasion when I was twenty-one and I was in for five years. I got into really good shape, but I tried not to meet too many friends. I didn't want any lifetime connections because they were the kind of connections that got me there in the first place. I just put my head down and did my job and learned about my culture."

There were other Cree and Blackfoot in prison; they had a Native Brotherhood society that formed to protect one another. They had weekly meetings. They'd talk and organize events. Cam worked making drums and tepees and bone chokers.

A few months into his employment with us I invited Cam to participate in a fund-raiser we were having on the farm. I had asked a few of our staff to get up and speak, and most agreed.

It has always been a question for us when and how and *if at all* our staff should be put before the public. Of course, some of our farmers have appeared in advertisements for Vancity, because of our relationship with them. And we're always encouraging our staff to work the farmers market stands. But in general we had stayed away from indulging in the "poverty porn" that we witnessed so many organizations use to raise money. We haven't wanted to parade our staff in front of the public like some sort of curiosities. But occasionally, when the time seems right, we ask them, because their stories are powerful, and without them it can be difficult to raise the funds to keep the project supported.

Our highest-yielding year to date was $300,000 in gross income from farm products sold, a couple of hundred thousand dollars short of our annual budget. Fund-raising has become one more in an ever-growing list of skills and jobs required to manage Sole Food. I have struggled with this piece of our puzzle. Although I have to remember that industrial agriculture is enabled to a great degree by huge subsidies and other hidden costs we all pay for such as polluted water, depleted soils, and poisoned food, again, as a farmer I want to be able to support our work by the food we grow.

I console myself with the knowledge that no farm can be competitive or survive if its attentions are as divided as ours. And I feel hopeful that as the skill level of our staff continues to improve—through training and employment and other supports—as they grow healthier and more stable, we might see our financial picture improve as well.

The day of the fund-raiser was spent feverishly cleaning up the False Creek farm. A large tent was set up in the middle of the farm; tables and tablecloths and chairs and sound systems all put into place. Several local chefs had volunteered and arrived with their food.

When all was set and the event began, several of our staff arrived, but drunk. Cam was one of them. At first I felt terribly disrespected by this. We had emphasized how important this event was, how we needed to raise funds to keep things going, and assumed that everyone would be on their best behavior. When the time came for staff to stand up and tell their story, it was a disaster. Two or three presentations in and Seann started frantically signaling me from the sidelines to cut it off—too many drunk rants. Cam was in the front row and had come prepared to speak, but I did not call on him.

At the end of the evening he stormed angrily out of the tent, upset that he had been asked to come but was excluded from presenting. Had he been sober it may have been different, but the program had already gone off the rails, and I had to put an end to it.

I felt bad that Cam was so upset with me. I tried unsuccessfully to find him the next day to talk it over. No luck. In the end I sent him a note. The

experience reminded us of the fine line we walk with our staff, between looking out for them and encouraging them to look out for themselves. And it reminded us that we can never know for sure what condition folks will be in when they arrive for work or to an event like this.

Cam told me that he sometimes catches himself singing Cree songs when he's harvesting. "I lose track and don't know how long I've been singing." His comment reminded me of my time with the Hopi in the early 1980s and an old Hopi farmer friend of mine, one who wore his wisdom so lightly. I remember this, something Little Dan told me many years ago: "Sometimes I come to my field in the evening and stay all night because the porcupines were eating my corn. I'd sing all the way up and down the rows. My dad said this corn is like children and you have to sing to it and then it will be happy."

Two roadways hang over the farm dominating the skyline to the north. The two parallel concrete behemoths were built in the 1970s in advance of a broader freeway system that would have plowed through and destroyed much of Vancouver's oldest neighborhoods, including Gastown, Strathcona, and Chinatown. In the end, protesters stopped the implementation of the broader freeway plan, and the success of Vancouver's community uprising was celebrated by urban communities all over North America. Still, the achievement came with losses too, as achievements usually do. The Georgia and Dunsmuir viaducts had already been constructed and the city's black neighborhood, Hogan's Alley, was demolished in anticipation of the proposed freeway construction.

Over forty years later the viaducts still carry traffic from downtown over the False Creek Flats, dumping it at the base of Main Street in Chinatown. But a new plan voted in by Vancouver's city council will demolish the structures and open up more greenspace. The change will reconnect the downtown core with the waters and remaining open land of False Creek.

Sole Food's False Creek farm will be a casualty of all this. Our site will likely become more condominiums adjacent to a new nine-acre park, an extension of Creekside Park that was promised to local residents twenty-five years ago. I love the idea of another city park. I support the need for more greenspace to soften the urban landscape and provide a place for people to gather. But many urban parks seem to follow the same boilerplate design, large expanses of grass, which requires fertilizer, water, and mowing, interspersed with a few trees.

I've had another idea. My vision for the new city park is a three-acre permanent farm space for Sole Food.

Imagine walkways winding through fruiting trees, vegetables, small livestock, and grains providing education, fresh food, and jobs for neighborhood residents. Consider the ecological, social, economic, and educational value of a farm within a park.

No longer living with the uncertainty of short-term leases, imagine what Sole Food could do with a permament location. In-ground plantings, perennial herbs and flowers, laying hens and bees, extensive orchards and other perennial crops, classroom space, a self-guided tour with plaques describing our farm philosophy and activities, a facility for canning, freezing, and drying perishable products for later sales and consumption.

I realize the enormous lobbying that will come from those who will want volleyball, or ice skating, or more soccer and baseball fields. My own son, if given the opportunity, would probably appear at public meetings in support of those uses over a farm. I am also well aware of

the push there will be for acres of grass so that families can picnic, play Frisbee, fly a kite, or just sit and look out over False Creek. But why could a creative design not include a combination of all of the above, in an integrated edible landscape?

Sports fields could be hedged with fruit and nut trees; hops and grapes could grow on the fences that tower behind homeplate or encircle the fields of play. Production vegetable fields could be intersected and surrounded with chamomile, mint, and lemon thyme "lawns," an olfactory invitation for parkgoers to lie down within the fragrance of those healing and relaxing herbs. Young people would have the opportunity to see how food is grown, witness the entertaining lives of laying hens, and participate and celebrate with the cycles of bloom, fruit-set, thinning, and harvest that an orchard provides.

An agricultural park would further one of Vancouver's top priorities, urban farming, in becoming one of the greenest cities in the world, and place the city in the forefront of creative urban planning and design.

While I would like to imagine a future without addiction, homelesness, poverty, or mental illness, it is not likely. Sadly I suspect that we will always need safe places for people who are on the edges to go, somewhere for those who are underserved to be valued and supported, work for people to do, fresh food to eat, and places where families and communities can see how food comes to them. Although our farming system, like many of our staff, is movable and transient, what we represent and the work we do must be rock-solid, constant, and deeply rooted. A permanent, central, accessible, and highly visible Sole Food home could provide that essential touchstone and be a model for cities all over the world.

Ritchie

CHAPTER 8

Arboreal Adventures

A fig catches my eye as I move down the rows of trees, my thoughts dominated by the need to check the sprinklers and water the orchard. I stop, though, and return to the branch where the fruit hangs. The fig is large, plump, and drooping low. From its base drips clear nectar. I reach out and gently curl my fingers around it. Then: that distinctive give, that soft lusciousness that confirms its ripeness. Ever so carefully, so as not to tear the flesh, I remove it, stem separating from mother branch, and take the first bite. Its sweet, gooey, red flesh envelopes my tongue, and for a moment I am transformed.

A police car barrels down the street, siren blaring, yanking me out of my fig ecstasy. I am back, standing on a one-acre lot at the corner of Main and Terminal, on one of the busiest intersections in the city of Vancouver.

The corner of Main and Terminal, when I first came to see this site, was not the kind of place where you'd want to hang out. Large trees had grown up through cracks in the broken pavement, and shattered grease-stained tile flooring that once supported a De Dutch fast-food restaurant and a Petro-Canada gas station, long since demolished. Once, this was a place where you could fill your stomach with hash and eggs or pancakes and your car with premium all in one spot.

Large chunks of concrete sat prominently in the center of the site in a sculptural urban Stonehenge accented by bent rebar protruding in all directions.

Makeshift, yurtlike hovels constructed using shipping pallets and plastic provided housing for a resident population of a dozen people who chose to live outside through the winter months, even though the

Aboriginal Friendship Centre shelter across the street offers one hundred people a warm space, a sleeping mat, and two meals a day from 5:30 PM to 10 AM.

When our Sole Food crew cleaned up this site, we needed a twenty-yard Dumpster to haul away the piles of old clothing and needles and trash that littered the lot. Some of that trash originated from the temporary resident population, but some of it—soda cans, empty bags of chips—had been tossed in by passersby.

It is here that we have established what has been referred to as the largest urban orchard in North America. I do not know the veracity of that claim, but I would bet that it is the largest urban orchard planted in containers. Like most of our urban farm sites, and most urban land in the world, the native soils here are contaminated. How this came to be is more than just the story of former gas stations and abandoned greasy spoons.

Prior to the arrival of Europeans, this part of Vancouver's False Creek neighborhood was a vast tidal mudflat surrounded by dense forests, all supporting a rich ecosystem teeming with bear, cougar, deer, elk, and migratory birds, and in the shallows and deeper waters a huge diversity of marine life. For thousands of years the Coast Salish peoples fished and hunted in these forests and waters. The first white settlers logged

and cleared the area, and by the late 1800s it had become a hub of industrial activity with numerous sawmills, shipyards, a salt refinery, foundries, metal fabricators, and a slaughterhouse. The development of the city of Vancouver—its roads, bridges, buildings, and much of its essential infrastructure—was fabricated here.

In 1910 the city of Vancouver gave the remaining 160 acres of that native marshland to the Canadian Pacific Railway for their yards and terminals. The last remnant of the rich tidal habitat that existed here, the final piece of a vanishing coastal ecosystem was about to be fully consumed.

The city's gift to the railway was based on an agreement with Canadian Pacific to fill the remainder of the marsh so that the land could be built on. No one knows for sure what was used as fill, but rumor has it that it was scrap lumber, bricks, industrial waste, and urban refuse. Within a few years the shallow waters and tidal flats were supplanted with warehouses, machine shops, and manufacturing plants. On the edges were numerous sewage outfalls, which still today discharge waste when heavy rainfalls overwhelm the city's sewage systems. The effluent ends up in False Creek, Burrard Inlet, and eventually the Georgia Strait.

Yet also today, a few hints of the wildness that vanished at the turn of the twentieth century are returning as Sole Food's trees provide a multistory habitat for birds, insects, and small mammals. And while it's possible to momentarily imagine that you are on some rural farm when you work in this orchard oasis among the towering cherry and quince and plum trees, the reminders of the urban surroundings are near-constant. Sirens and traffic, SkyTrains passing overhead, new office and condominium buildings that seem to be straining and stretching upward faster even than our herbs and trees.

There is history and movement under our feet here at Main and Terminal. We're farming atop a visible record of urbanization, and our orchard is only the most recent layer of development in this part of the city. And all along, despite every effort to subdue it, nature (and our commitment to it) has continued to be a force.

In 2013, when a city construction crew excavated a huge hole next to the farm to run a new sewer line, the remains of what looked like train trestles or old wooden shipping docks were revealed. We watched in awe the power of nature, unstoppable even under major urban streets, as the tide came in and filled the gaping hole in the middle of the four-lane Terminal Avenue. You can fill and pave over and think you have conquered, but the power of the sea cannot be stopped. Rumor has it that close to a million dollars was dumped into that hole before the city postponed the project for safety reasons.

Signals

When folks find out that I am an organic farmer, the first question they ask is, "What do you do to control the pests?" This surprises me—it's as if they view all farming as some sort of heavily defensive act, that farmers are like generals in the field fighting off an endless invading force. My answer to the question is almost always received with some level of doubt and suspicion.

I tell people that pests and diseases are the result of a system out of balance. It's no different for you or for me. If I push myself hard, don't sleep, and don't eat well, eventually I will get sick. Just like people, plants require a well-balanced environment to stay well. Plants under stress attract pests and diseases.

The best pest and disease strategy is to create the conditions for dynamic plant health. This starts by building and nurturing soils so that they are well balanced nutritionally and biologically and by providing growing conditions that are consistent and stable.

But even within the most well-balanced biological system environmental or climatic conditions are such that pest or disease pressures can occur, and although an arsenal of biological and botanical materials is available for organic growers, we rarely spray anything unless the problem has reached a significant economic threshold. If the crop is so valuable that losing it means the farm may not be able to pay its bills, and if the pest or disease pressure is at a level that it will ruin most of that crop, then I will interfere. Otherwise the interference, the cost of the biological or botanical control material, the impact on the ecological system of the farm are not worth it. We will let that crop go.

It is rare that this happens or that we ever have to implement any insect or disease controls. Our primary challenges are related to rodents, deer, birds, and, in some cases, humans. The best and least invasive strategy for these pests is exclusion, row covers, screening, fencing, and, for rodents, simple trapping.

The large yellow neon PACIFIC RAILWAY sign on the entrance to the 1917 classic Beaux Arts–style building that houses Vancouver's main passenger rail station is visible from one corner of our lot. So are the 155-foot-tall geodesic dome of Science World; the newly constructed fourteen-story Vancity credit union head office, which has trains running through its center; and the new Science World commuter-line station. Farther down Main Street and directly across from the farm is a

McDonald's restaurant, a constant reminder of the conflicting worlds we live in, and a cluster of new condominiums. On the other side of the farm, on Western Street, are the remains of some of Vancouver's oldest standing brick and now rusting corrugated-metal industrial warehouses. These are home to the Aboriginal shelter, a hotel-furniture liquidator, a tile shop, a boarded-up storage space, a welding shop, and a Fiat dealership dotted with forest-green, pearl, mocha latte, and blue tornado Minis planted at the same density as our trees.

This is a neighborhood in transition, and it seems it always has been. A high-rise condominium started a year ago leapt out of the ground with amazing speed, and plans are under way to demolish the remaining derelict and graffitied neighborhood buildings that flank us on Western Street.

This urban tumult, the constant building up and tearing down, witnessing architectural and industrial visions come to life all around, inspired me to want to create something that had the qualities of a forest, provided shade and refuge, and made fruit. I imagined a visually striking and movable perennial orchard, made up of diverse species of fruit trees. Our trees could provide habitat for birds and beneficial insects and supply food year after year without the constant churning of soil, replanting, and fertility inputs required for vegetables.

As I've said, everything I do begins with a picture in my mind. This vision started almost forty years ago when I first started seeing abandoned lots in decaying neighborhoods in America's cities and began

imagining how they could be transformed—and, in turn, transform the communities that surrounded them. My early manifestations of this dream involved annual vegetables, and in Vancouver that vision became a reality first, even as I fantasized about a bountiful urban orchard.

It's true—and Sole Food bears this out—I love carrots and beans and melons and greens. I know they are what most people picture when they think *farm*, especially urban farm. But even as we planted fast-growing vegetables that first season under the Astoria in order to pay the bills, I had in mind that within the urban context an orchard may be far more practical. Lateral space is limited in cities, and it made good sense to consider a farming system that used vertical space and could keep on giving without having to till soil every couple of months or haul in vast amounts of nutrients. I could see the beauty of an orchard, the shade it would generate, the conversion of the urban heat sink into edible calories, the gathering place one could create.

Everyone should take Lyle's homeboy walking tour of the neighborhood. The tour Lyle gave me one fall day in 2014 did not feature stops at scenic outlooks, museums, architectural wonders, or sacred sites. His private tour was into the back alleys, the dive hotels, the vending machines offering crack pipes for twenty-five cents, the parks, a revival meeting, and a soup kitchen. He took me to his own home, in the Station Street Community housing project, where upon our arrival residents were lining up for an experimental program that provides regular measured prescription doses of beer and wine, a way to help alcoholics stay away from Listerine, hand sanitizer, and other toxic sources of alcohol.

"Who are you, the fucking mayor or what?" I asked Lyle as he stopped to talk with almost everyone he encountered while we made our way through the streets and alleys. Along the way he filled me in on the history of certain buildings, including Insite, the safe injection facility that he claims saved his life, and the personal stories of folks we encountered along the way. We walked by a hotel and he described the gruesome scene of someone tossed out of a window during a drug deal gone bad.

Along the way he stopped, posed, and asked me to photograph him in his muscle shirt in front of the spray-painted sign that says KEEP OUR CITY CLEAN, DEPOSIT USED RIGS HERE.

Lyle joined Sole Food in 2013, after working at the recycling depot, providing "heavyweight" collection services for unpaid drug deals, selling heroin, and participating in petty theft.

Everyone in this community knows Lyle, and he carries himself through this underworld with confidence and a sense of pride one would not normally associate with such a place. There is community here in the real sense of the word; people are out interacting, exchanging, hanging out in groups. The day we walked the neighborhood, the streets were relatively quiet. It was Welfare Wednesday, people here had gotten their checks, and most were holed up in their rooms getting high, or so said Lyle. Even so, there were more people on the streets here than in the more upscale neighborhoods I'd passed through earlier on my bike.

I listened to bits of conversations as we moved through the neighborhood. Lyle checked in with a guy pushing a shopping cart who was collecting cans and bottles, asking about his health. He stopped to commiserate with a street nurse who described the number of funerals that have been taking place lately. I noticed random acts of generosity in these streets too: a cigarette passed on to a stranger, a pat on the back, a dopesick junkie getting help from a friend. Lyle revealed one of the ironies of life here, and probably in much of the world: Those who have the least are often the most generous.

Despite my work with Sole Food, my relationship with this neighborhood at large remains peripheral. I've walked and biked and cruised around here numerous times over the last seven years, and even though very little shocks me these days, I still feel like an outsider. Touring with Lyle, I could just as well have been in some foreign country where I didn't speak the language, didn't look or dress in native style, and was dependent on a local to show me around. With Lyle, I felt like a gawking tourist who had decided it would be cool to return home with some tales and photographs of his visit to skid row.

How do I portray the reality of the scene—of Lyle's home here—without being sensational, voyeuristic, or taking advantage? This question dogs me. Some people turned their heads away when I was photographing, while others approached me and wanted to be photographed. I've been making photographs my whole adult life in all kinds of situations and in every part of the world, but here I felt awkward and hesitant. I stopped long enough to make one image of a cluster of people hunched over in an alley, used needles on a chunk of concrete next to them, and a body sprawled on the pavement nearby.

Lyle is as comfortable here on these streets and alleys as I am not, but he's got the credentials. He leaned in as we moved on. He had a story to tell. "In 2002 I jumped out of an eighteen-story building and hit the fifteenth-story balcony on the way down," he said. "They put me in a coma for three weeks, and I was in the hospital for nine months. I lost my family and my kids. Then I went homeless for about nine years.

"I got issues," he continued. "I'm addicted to heroin, and I need 250 milligrams of morphine every day." In his fall, Lyle broke every bone in his feet, which were then bound for a year. "I was supposed to get an operation but just went into my addiction instead."

Lyle told me that he'd lived a block and a half away from one of our farms but didn't notice it at first. "Most people don't see this place," he told me. But once it appeared on his radar, he decided to step in. "I was on to it like hotcakes. I love the people I work with at Sole Food. They're not judgmental. If you call in and can't work you're not going to get fired. You don't have the weight of the job beating down on you."

Lyle's right, I think, when he says that most people don't want the addicts. And yet he believes—and he's convinced me—that around our orchard and in the streets connecting our farms is a sense of belonging. It comes through the work as much as his daily walks through the community. Lyle is equal parts mayor and farmer.

"My favorite part of the job," he said as we returned to the farm, where I'd chained up my bike, "is turning the soil over. It's therapeutic.

Starting things from scratch and seeing them grow is amazing. It's strange that my punishment as a kid was weeding the garden."

———————

In the early 1980s I joined a commune in Southern California that was based on agrarian principles. We had three different parcels of land totaling some four thousand acres on which we raised row crops and tended orchards, operated a full-scale goat and cow dairy, and produced grain and fiber. We supplied our own natural food stores and bakery and juice factory and restaurant. We fed ourselves. We even made our own clothing, backpacks, and shoes.

After only four months living in that community I was given the responsibility of managing the hundred-acre pear and apple orchard located in a high desert valley east of Ojai, California. At the time this was one of only a handful of commercial orchards in the country that was farmed organically. Here I was at the age of eighteen with no orcharding experience, having never managed anything, directing a crew of thirty people, most of whom were older than I.

The orchard had been abandoned for fifteen years; the branches between trees had become so intertwined that you couldn't find the alleys down the middles of the rows.

For guidance, I had a 1930s copy of *Modern Fruit Science* and the journal of the guy who ran the place the year before, who'd given up in frustration. Even way back then, I kept a copy of that Goethe quote about boldness and genius and magic attached to the door of my twenty-foot unheated trailer.

This experiment could have ended really badly, and had it failed—and maybe it should have—I would have probably spent the rest of my life working in a high-rise office building. But there was something that took place down those rows of apple and pear trees, something very different from what is happening in most agricultural fields and orchards in North America.

I went to work each day with thirty of my friends, and while we worked we joked and talked, and we discussed our dreams. We tried out our latest theories and philosophies on one another, speculated on the fate of the earth, and ate our lunch together in the shade of the trees. In the winter we pruned every day for four months straight, in the spring we thinned fruit, and in the fall we ran the ten-week harvest marathon.

It was repetitive work. But at the end of each day, instead of feeling I had been chained to mind-numbing drudgery, I felt like I had attended an all-day party. The orchard thrived, and those apples and pears gained

a reputation around the country. And while the cold nights and hot days of that high desert provided ideal growing conditions, I am sure that fruit was equally infused with the energy of that group of people and the pleasure we found in one another and in that land.

This community orchard experience was my introduction to agriculture, and it has informed all of my agricultural endeavors since. It demonstrated that good food is more than just the confluence of technique and fertile soil; that it is the result of men and women who love their land, and who bring passion to working with it. It's extraordinary to me that here in the city, someone like Lyle, even in the throes of addiction, can experience that same passion.

That orchard from my past provided the place where we could work together as a community every day, and the cycles of those trees became the cycles of our lives, the rhythm of our year, a structure, and something predictable in a world that was otherwise not.

There are many communities around the world where lives revolve around orchards and the fruit they produce. In Italy I visited five-hundred-year-old olive groves, witnessed ancient olive and carob trees wrapped around each other in a primordial embrace, found communities whose central focus was the harvest and the pressing of oil and the cultural cycles, celebrations, and rituals that went with it. I've witnessed this continuity and sense of community all over the world, in the ancient apple orchards of England, the craggy overgrown mango groves in Jamaica, the former Gravenstein orchards in Sonoma, California, and the apricot and cherry orchards that used to cover what is now Silicon Valley.

Every plant and all food has a story. Knowing that story not only enriches our experience as growers and eaters but also enhances our knowledge of how to grow and prepare those foods.

The story of fruit is the story of culture. Many of the fruits we commonly use evolved with the cultures that propagated and grew them. Think of figs or olives and you think of Italy and Greece and the region around the Mediterranean Sea. What is the climate like there? The culture? How much does it rain and when? What are the soils like in those regions?

This knowledge informs our experience and makes us better farmers, whether we are growing figs or spinach or strawberries or lettuce. It also reminds us that we are just one link in a long and rich cultural and ecological chain. Our relationships with plants are similar to our relationships with people: Each has a particular personality, each requires being present, paying attention, being available, taking care. Yet to me it seems more possible to make friends with fruit trees than with a carrot or tomato. Trees are simply around a lot longer. There's more time to build a relationship.

ARBOREAL ADVENTURES

I spent half of my adult life on that Southern California farm anchored by fruit trees: peaches, apricots, nectarines, plums, Algerian mandarins, limes, cherimoyas, pineapple guavas, navel oranges, and avocados. The vegetables were grown down the alleys of the trees in a beautiful, highly productive polyculture.

I planted most of those trees, but the avocado orchard had been established long before I arrived. I heard that it was first planted the year and the month I was born. I referred to the avocado orchard as "the cathedral." The trees stood forty to fifty feet tall and formed a forest canopy. Entering that space was like entering a place of worship; it was quiet, dark, shady, and cool even on a hot day. Look up, and there was a lush green canopy high above with clusters of pear-shaped green pebbly fruit hanging throughout.

Fifty years of leaf-drop created soils in that grove that sustained the avocado-forest ecosystem. Every tour I led on that farm culminated in the avocado cathedral. We would sit down in a circle, and I would guide the group through a journey into the world of the soil. The floor of that orchard was so rich and biologically active that everyone who inspected and smelled that soil instinctively knew that there was magic in and beneath those trees.

———————

Establishing an orchard on a derelict lot in the middle of the city is a challenge. Even more so for the Sole Food orchard, because we planted

the trees, each expected to live for twenty to twenty-five years, with their roots in a box, not in the ground. Would the roots of the trees begin to suffer once they filled the capacity of the box? Could we prune the tops to compensate for the limited root space? How could we maintain the biology and soil fertility in containers with a long-term perennial crop? How could we create enough drainage and maintain just the right level of soil moisture so that we're not using too much expensive municipal water?

We also had to consider how to dissipate heat growing fruit trees in containers. The black plastic containers become solar collectors. This can be beneficial, because heat dramatically enhances growth, fruit production, and fruit quality. But some species suffer if their roots become too hot. Figs, citrus, and persimmons want the heat; many of the other deciduous trees such as cherries, apples, plums, and pears seem to have their limits. Cherries seem to be most sensitive.

Planning and planting this orchard was just one more experience of jumping off the cliff without knowing what I would find on the way down. This willingness to take chances, to experiment, has fueled the great agricultural experiment for these thousands of years. How much experience do we really have with the unique aspects of urban farming?

Building on the knowledge we'd gained from growing vegetables in large containers, we redesigned our boxes to be stackable, nestable, and fully movable, with forklift tabs under each one. We had them manufactured out of plastic rather than wood. We drilled holes around the bottom to allow for drainage and placed gravel in the base of the boxes.

Portability was critical because we were installing the orchard on leased land. If a lease could not be renewed at some point in the future we would have to move the trees, and we wanted to be prepared for that. In fact, the portability of the containers was tested soon after we established the orchard at Main and Terminal when the city announced that they would have to bring a giant excavator through the site to install a sewer line for the new condominiums being built nearby.

It was a major inconvenience, but it proved a great opportunity to test our system. Would the boxes hold up to being moved? If we bent a box while moving it, would the soil shift and open up, disrupting the root systems of the tree? How much time would it take to move all these trees?

Using a set of forks attached to a Bobcat tractor, we moved the trees out of the path of the excavator quickly and efficiently, without any mishaps. The trees did not notice they had been moved either; they continued to grow and flourish in their temporary location as if nothing had happened.

Simple Beauty

Asked why I farm, I like to say so I can eat well and feed my family and community, so I'll have stories to tell, and so I can keep my sanity. But I also farm for the simple beauty of it. We tend to think of farms as utilitarian food factories, but whether rural or urban, farms can be places of incredible beauty. Herbs planted as an understory below fruiting trees are incredibly productive. They have provided a diverse orchard ecosystem, but they also fulfill a powerful aesthetic role.

In peak summer when you pass through the orchard gate at the corner of Main and Terminal out of the noise, away from the traffic and exhaust and constant assault of urban hum, you can walk down the rows of cherry, plum, quince, apple, pear, and persimmon canopied overhead and witness the display of flowering chives and sage and oregano and mint, and you cannot help but be transformed. Harvest a few leaves, crush them and inhale their essence; the experience overwhelms your senses.

Figs and quince and lemons were the first to bear fruit, but now plums and apples and pears and cherries have followed. To stand and eat from these trees, on what was only a few years ago a contaminated and abandoned piece of the urban wilderness, is amazing.

It is a productive place, this oasis of fruit, flower, and herb, but it is also a refuge, a place to rest your eyes from hardscape gray and black to restful green, someplace to find shade from the heat of the city, and to sample a leaf of mint, a yellow plum, or an apple or pear.

When I worry about the potential constricted container space of our urban orchard, I remind myself about one of my basic understandings about growing fruit. Whether it is strawberries, or figs, or melons, or peaches, the ones that taste the best are almost always those that have suffered. Having everything you need all the time, living a life where the road is straight and smooth and without challenge, does not develop character and depth. Those misshapen, sometimes smaller strawberries or apricots inevitably have the most sugar and intensity and complexity of flavor.

Grape growers understand this principle well: Hold back the water, provide a little stress, and the sugars and flavors concentrate. In California I dry-farmed our tomato crops. I let them grow as tall, lanky plants in our propagation nursery and, in the spring, planted them eighteen inches deep in the field, where they never received a drop of irrigation. The result was always tomatoes with rich flavor and sugars. Those plants suffered, but the fruit was good, really good.

While farming in California, I used to take hundreds of pounds of our seconds peaches, fruit that was bruised or cosmetically imperfect or misshapen, to dry in the sun on an old apricot ranch in the Upper Ojai Valley. The climate was not supportive for drying on the coast where I farmed, and the Halls' apricot ranch had a cutting shed with hundred-year-old wooden drying racks, a mini rail system for moving the racks full of cut fruit into the field, and daytime temperatures that would peak above one hundred degrees Fahrenheit. When placed out in those fields, the fruit would sear over instantly, and in a few days a large peach half would be fully dry and delicious.

I suspect that the apricot trees at the Halls' ranch had never seen a drop of irrigation water. The trees were rough and craggy and old. The fruits they bore were small and not very pretty, but unbelievably tasty. Eating them was like consuming small globes of pure sunshine.

Is it possible that the trees we are growing at the corner of Main and Terminal, trees that will surely face adversity in constricted containers, in the harsh environment of the city, in a climate that is marginal, will grow up to make babies that will be so tasty they'll knock our socks off?

When I started this orchard experiment I was accused of having a deep-seated addiciton to gambling, not just because we were growing trees in boxes or in the heart of a big city, but because of my choice of species. Who ever heard of growing citrus in Vancouver or persimmons or quince or even figs?

And although it's true that farming is the most prevalent form of legalized gambling, I am not a gambler. I'm not betting on global warming to see my experiments through; I'm not taking chances that are entirely baseless or without some deep thought or research. Just because something has not been done does not mean that it cannot be done.

The grafted bare-root trees that arrived early in the spring of 2013 had to be kept in the walk-in cooler of a local restaurant until we completed lining out the boxes in long straight rows and filling them with soil.

We scheduled a work party to fill the boxes with soil, but after a full day with fifty people working, hands and bodies and wheelbarrows and shovels moving as fast as they could, we realized that completing this job by hand would take months. So we rented a John Deere crawler loader, which freed up most of the crew to keep up with planting and projects at the other farms. Two crew members measured and placed boxes in rows while I maneuvered the tractor in and around the tight spaces to fill each box.

Filling eight hundred boxes to complete the orchard site was slow work. Even with the added power of the machine, we were over a month behind schedule, and I could almost hear those living trees in the restaurant cooler, screaming to be planted. I knew that because of the extreme delay, some of the trees would likely never take root and would die.

It was almost July when all was ready, months later than bare-root trees should ever be put in the ground. We marked out the rows, keeping like with like, and began placing each tree in its assigned box.

The planting progressed fast. I've participated in setting up many orchards over the years, all on open land, but I've never seen trees planted so quickly. The soil was so loose, and there was no need to stoop over while planting—both those factors accelerated the process dramatically.

Within a week, the trees were pushing out new buds and within several more weeks some of them looked as though they had been in the ground growing for months. At the end of the first season, after only four months of growth, the orchard looked like it was several years old. Those raised black boxes filled with highly fertile, loose soil and placed on warm pavement in the city provided everything those trees needed. They responded with gusto.

The Miracle of Persimmons

I have always advised beginning farmers, "Grow those crops that you like to eat; you'll do a better job." The persimmon requires that I amend that advice.

In all honesty, I don't particularly like the taste and texture of persimmons. It's not that I have a bad memory of having my mouth painfully puckered from eating an unripe one, or that I've never had a well-grown one. I'll eat them dried or frozen but rarely would I choose to eat one fresh. I love this fruit and the trees that produce them for other reasons.

The tree is beautiful at every stage, from first leaves, flowers, and new-set fruit to the falling of autumn's burnt-orange and red foliage, which reveals fat Day-Glo globes hanging like ornaments on leafless trees. Whether it is the large, plump, and pointed Hachiya, or the flat segmented Fuyu,

the persimmon is such a pleasure just to look at, no need to touch or sample.

I've heard the stories of perfect persimmons, harvested with stem and a few leaves, purchased in Japan for $30 or even $50 each to be given as gifts, part of a tradition that now includes square and heart-shaped watermelons, $6 strawberries, and Buddha-shaped pears. I so appreciate that culture's love of fruit. In Japan fruit is not just something to grab and inhale, but food that ought to be enjoyed slowly, visually, aromatically, and only finally gastronomically.

My amended advice: Grow those crops you like to eat, but also grow those that you simply like to look at.

Cultivated for thousands of years, the Asian persimmon originated in China and includes hundreds of cultivars. The botanical name for this fruit is *Diospyros*, which translates to "food of the gods."

Yet the following spring the consequences of the prior season's late planting were apparent. Twenty-five trees had not survived. This does not sound like a lot, but it is rare to lose orchard stock, the young trees are expensive, and the time lost in replanting and waiting for new plantings to catch up can be significant.

Alain and his son Dion helped me remove the dead trees and replant. I always enjoy watching Alain working with Dion, who for all his size and strength will always be part child. I'm in awe of Alain's patience and kindness with Dion.

I assigned them the job of removing the dead trees. I showed them how to scratch with a fingernail to check for any sign of green life beneath

The persimmon is typically grown in subtropical climates, but I planted twenty-five trees in our Sole Food orchard. And while our Vancouver climate is considered to be relatively moderate, it is definitely not subtropical.

There was doubt, hand-wringing, and even consternation when I planted persimmons, figs, and lemons in that orchard alongside the more typical apples, pears, cherries, and plums. Our funders, neighbors, and even some of our staff thought that maybe I had gone too far.

And so I felt a defining moment when in the fall of 2015 we harvested over four hundred pounds of these fruits from trees that were barely three years old. There were still naysayers; a few farmers who saw our persimmons in the kitchen of a local restaurant were heard suggesting that we had "purchased" them in Chinatown and "repacked and sold them."

Jealousy and competition in the farming community can be a wonderful thing; it stimulates fresh discoveries, innovations, and new crops. I expect to hear that those same farmers are now planting persimmon trees on their land.

But in truth that first bumper crop of persimmons may be an anomaly. The summer that bore them was unusually hot and dry, and one year's harvest is not proof of lasting success.

Even so, after that first harvest I delivered several of these fruits to the mayor and city manager's office, in part as a thank-you for their friendship and support, in part to demonstrate what is possible in the city, and on some subtle level as a warning that the presence of these fruits, while incredibly seductive, may be a sign that climate change is well under way.

During our year-end staff gathering I held up a Sole Food persimmon and referred to it as a miracle, one of many we have witnessed on our farms.

the brown bark. I demonstrated how easily a tree that had died could be lifted out of its container—the roots would not put up any resistance.

Alain and Dion went off to work at the other end of the orchard. Half an hour later the duo came walking up to me smiling and proudly holding a live fig tree with its roots dangling, commenting on how difficult it had been to pull it out.

It takes incredible effort to dislodge a living tree like that. And yet there it was—completely uprooted!

We acknowledged the error and had a good laugh. I realized I could have done a better job explaining things, and Alain now fully understood the difference between a tree that is alive but still sleeping and one that

has perished. In that moment we were both reminded of the value of mistakes—hopefully ones that are not too expensive—to inform our experience and growth.

Later that day, Dion told me he was happy to be out working, that when he was at Sole Food he was "just tryin' to learn how to behave." I told him I was trying to do the very same thing.

————————

An orchard has an energy unlike any other agricultural endeavor. Fruit trees can grow to be very old, in some cases representing hundreds of years of relationship with human communities. The cycle from flower to fruit is long and drawn out, one that allows for contemplation, community participation, and celebration.

While the products from an orchard can be highly nutritious, tree fruit is often about pure sweet pleasure, a way to begin the day or end a meal. Fruit can be eaten fresh, sauced, made into pies or crisps, frozen or juiced or dried for yearlong consumption. Working in a field of row crops involves bending over or squatting, looking down and deep and detailed. Working in an orchard demands that we look up, climb, and reach.

Main and Terminal was about as unlikely a place to fulfill my vision for an orchard as I could imagine. And yet this site now contains producing persimmons, figs, apples, cherries, plums, pears, and quince. Close to five hundred trees are thriving and making fruit on that site.

We planted our trees in every other box along the rows in order to give them the space they would require as they developed. In the alternate boxes are an array of culinary herbs: chocolate mint, spearmint, peppermint, lemon mint, savory, lovage, sage, lemon thyme, marjoram, Greek and Italian oreganos, lemon balm, chives, and more.

Our urban orchard is now well established, the trees have trunks four and five inches in diameter, and the herbs continue to fill the space between the trees and spill over and down the sides of the containers. On a summer day the hum of insect activity is intense; the bees and wasps and flies almost overwhelm the din of cars and bicycles and pedestrians that buzz around the edges of the site.

The orchard feels complete, like a little self-contained dynamic world, the tree branches now reaching over the narrow pathways between the rows, quince touching apple, plum and pear and cherry intertwining, fig and persimmon limbs laden, hovering over black asphalt paving.

If you're walking on the sidewalk along the edges of this space, the allure of fruit and shade and dynamic plant life is real. It's inviting. The temptation to climb the fence can be overwhelming.

In the fall of 2015 close to two hundred pounds of apples—probably more than we should have allowed those young trees to support—disappeared in a single night. The thieves left behind the handmade extension tools they used to harvest the fruit. The devices were impressive in their ingenuity, as was the thieves' thoroughness.

Weeks before this incident I arrived at the orchard and found an elderly homeless man harvesting herbs from below the trees. When I asked him what he was doing he told me this was "God's food" and ignored me and kept on cutting. I responded that God had a lot of help from our crew in planting and nurturing those herbs and that we needed to sell them in order to provide paychecks and a place to come to learn and work. I told him he could keep what he had harvested and that we are always willing to share if we are asked first, and that he should leave the site.

This exchange reminded me of an ongoing internal conflict. Deep down and apart from the farm production mentality that often gets in my way, I like the idea of opening up this orchard for free-form Garden of Eden consumption and enjoyment. I struggle with the idea that this or any other of our farms are cut off, protected, and inaccessible. If you spend your life on the street, and you're hungry or harried, hassled or tired from the stress of trying to survive, you could let go here, and immerse yourself in the magic and healing and pure edible sweetness of this little oasis. You could allow it to wrap itself around you and infuse you with its aromas, its sweet flavors, its safety and shelter, and its simple beauty.

CHAPTER 9

Occupy

Theft is a fact of life in the neighborhood where we work. Anything that is not bolted down, securely locked, or well guarded will disappear. It is nothing personal; when someone is dopesick or needs a fix he'll do whatever is necessary.

Four scales, five new market tents, backpacks, a circular saw, six drills, the farmers market cashbox, a Sawzall, jugs of sanitizers from the portable toilets (to drink because they contain alcohol), clothing, boots, screws, nails, bolts, wheelbarrows, bicycles, aluminum greenhouse parts, computers, live electrical wiring, an entire truck-cooling system, and more have been stolen from our farms. The skill required to pull off some of these heists is impressive.

And while our farms are especially vulnerable and difficult to fully secure, just working, living, or having a cup of coffee in the Downtown Eastside makes you fair game. One afternoon, I am sitting with Seann and Tabitha in the back of a coffee shop not too far from our False Creek farm. I have my Apple laptop on the table, plugged in, my hands on the keys typing. Just then, a man walks in, sidles up to the table, rips the computer from under my hands, and heads for the entrance.

He stumbles and drops the computer, which bounces off the concrete floor; he grabs it again and heads for the door. I recover from the initial shock and figure out what's happened. I'm up and out the door behind him. I chase him down the middle of the street, flying. *Who knew I could run this fast!* There is no way he was going to get away.

I can see the thief cannot believe that this guy twice his age and half his size is actually chasing him and has caught up. We're dodging cars. I grab his shirt, but he breaks away. One block, then two, then on the third block I catch him again. I latch onto his arm, but he's too big for me to pull down.

We're on the sidewalk on a corner now, and I'm struggling to wrestle him to the ground. Over my shoulder, I glimpse Seann barreling down the block. Seann is a former hockey player, and he's in the air, slamming into the guy. They both go down with a thud onto a steel grate with the computer beneath them. There is a scuffle. Seann holds the guy face-down with his arm behind his back.

I pull the computer from under them and call the cops, while the thief pleads with me not to press charges. I know I'm not going to do that; I just want to give him a little scare. Had he gotten away with the computer he would have probably sold it for a hundred bucks, enough to buy a few hits of crack or doses of heroin. It would cost the legal system thousands to haul him to jail, house him, feed him, and prosecute him—and he'd be out on the streets doing the same thing within weeks. In the minutes it takes for the police to arrive I let him suffer.

A crowd has formed. Someone yells out, "Does the computer still work?" *There is no way*, I think, having watched it bounce off concrete, then smash between two men and a steel grate and sidewalk. "Turn it on," someone else yells. I open the screen and watch in amazement as it flickers to life, functioning perfectly.

Seann and I walk back to the coffee shop to grab our stuff. We receive applause from the customers who witnessed the scene. I make a joke to the crowd that this was actually a commercial for Apple and head out the door.

I used that computer for another year with no problems and then passed it on to my eldest son who, within a week, had it stolen from his car.

We have placed steel shipping containers on each of our farm sites, and we use them to store tools and equipment and as offices and refrigerated storage. It's difficult to break into one of these containers; they have recessed, thick steel lockboxes, and the walls are made with fourteen-gauge welded corrugated steel. Even so, someone cut a hole in the steel mesh that covers the single window of our farm office container just large enough for an arm to reach through and grab a Sawzall. Apparently the thief did not have enough time to make a hole large enough to climb through.

The hoop houses—steel frames covered with plastic that we installed to grow tomatoes, peppers, melons, eggplant, and basil—were almost complete when we received a stop-work order from the city. It was several weeks later when we resumed work, this time with a building permit in place. We gathered all our tools, organized who would do what, and started work, only to discover that all the aluminum stripping required to finish the job was gone. Five thousand dollars' worth of aluminum greenhouse parts had been stolen in the weeks that the job was held up. Seann checked at all the metal recycling centers in the neighborhood and sent one of our staff down Hastings Street in the hope of finding the parts for sale, but no luck.

The live copper wires, which carry 120 volts of electricity, that service our inflation and circulation fans in those same greenhouses were pulled directly out of the conduit with their bare frayed ends left dangling.

Seann parked our delivery step van at that same farm for one night, returned the next day, started it up, and drove it back to False Creek. I was waiting for him that morning when he returned to report there was something wrong with the truck. When I opened the hood I discovered that the entire cooling system—water pump, radiator, everything—was gone. I still cannot understand how he was able to drive it several miles without the engine seizing up.

There are times when we have walked down Hastings Street after a theft and found a stolen item for sale on someone's blanket along the sidewalk. Normally we just let it go, knowing that in the end it is our responsibility to keep the farms and our tools secure. It is a game we play, always trying to stay one step ahead of the next theft, creating strategies and security measures that we can afford and that we hope will work.

But there is a limit. We will never have the resources

to fully secure all the sites all the time. As long as there are folks who have drug habits, as long as our society is divided into haves and have-nots, things are going to disappear. I sometimes naively think that the best security is our community-based employment, getting our story out, our staff communicating to their peers. But when you desperately need something to eat, or when you need a fix, reason and moral imperatives tend to fade out.

Although theft is a problem we may never solve, rarely is the food that we grow stolen. (The orchard thieves were a real shock.) Still, I would rather see kale and chard, tomatoes and strawberries, melons and peppers disappear from our farm sites, knowing that there is some nutritional benefit being passed on. One day, I hope, a box of fresh food might be in as much demand as a wrench or a piece of copper wire, a radiator, or a computer.

———————

There are places in every community where essential systems that keep things functioning are stored, warehoused, maintained, and supported. Buses and taxis need a barn to return to in order to be groomed and fed and serviced; school districts need a place to store desks and the tools and equipment to maintain buildings; and sanitation crews need a repository for sorting, processing, recycling, or disposing of all the detritus of a populated place. Although we all depend on these services, they are usually situated out of sight and out of mind. We flush the toilet, and our waste magically disappears; we put out the garbage in the morning, and the cans are empty by afternoon; we go to the store, and food is always waiting on the shelf.

Every city has its support areas: its storage lots, warehouses, scrap-yards, and transfer stations, places where sand and cement and lumber and steel and pipe are stored and organized for distribution. These are the areas where trucks, trains, and ships can load and unload and con-solidate and deliver.

Our one-acre farm at 1580 Vernon Drive spent part of its life as a gas station in an industrial part of Vancouver on the western end of the Grandview-Woodland neighborhood. The farm is directly beneath the First Avenue Overpass down the street from Yellow Cab's mainte-nance yard and parking depot, Davis Trading metal scrapyard, Ruff Stuff Dog Services, Unitow's high-security lot for misbehaved and impounded vehicles, a rail yard, Able Auctions, and General Paint's full-block warehouse.

OCCUPY

In the 1930s, triggered by the Depression and the extended drought in the Canadian prairies, young men migrated en masse into Vancouver, many settling in makeshift encampments in this neighborhood. The Grandview area at that time was referred to as the "backdoor" of the city, with one local official commenting that it was "an unpleasant sort of place inhabited by an unpleasant sort of people," referring to another essential but hidden element of city life—the working class.

During the months when Seann and I were cruising the city looking for ideal farming sites, this site we call Vernon is the one that I immediately liked. It was flat and half paved, already fenced, and located within a neighborhood that was undervalued. There were large holes in the fence that individuals would slip through at night to unpackage stolen goods or find refuge in the overgrown brush along the edges or in and among the steel girders and cement supports that hold up the bridge dominating one side of the site.

It took about a year from our first view of the site until the city of Vancouver offered it to us. Another year elapsed as we navigated the environmental, physical, and bureaucratic maze. We persevered, though, and Vernon has become the home for sixteen thousand square feet of unheated tunnel houses, the place where we can maximize the essential requirement to extend the seasons and produce heat-loving and cold-sensitive crops both early and late.

In the end, though, the ups and downs of our northern climate have proved to be far less of a challenge than the financial stress created by permit requirements, environmental consultants, utility installation, and the major losses we've experienced as a result of theft and vandalism by those who frequent the site after the sun goes down and darkness provides protection and cover.

————————

Mike Divine came to us through Mission Impossible, a Christian organization on the Downtown Eastside that helps people find work. Following the series of thefts at the farm, topped by the live copper wiring pulled from its conduit, we concluded that we needed someone to provide nighttime security. Mike took the job.

Mike and I met one Sunday at the Vernon farm to talk. He told me how he had been getting ready to retire: His kids had grown up, and he had put away enough money. It was all set. But then his wife left him, and the tax department came knocking. He lost everything, got into drugs, and ended up on the streets. Mike's story features a thread that's common in

Mike

this neighborhood: One tragedy, one major loss, a divorce, and everything unravels.

Mike entered a long-term recovery program, along with nine other people. He was the only one who made it through. He attributes his success to the fact that he was older than the others.

While doing his nighttime security rounds for Sole Food, Mike has come in contact with a cast of characters. On one of his first nights out he encountered a guy crouched down behind one of the greenhouses, stripping wire from its plastic coating. On being discovered the guy threatened Mike with a screwdriver, only calming down after Mike described the purpose of our farm. Another guy snuck up on him, trained the beam of his flashlight on Mike's face, and interrogated him. And there were individuals like Jim, who had made his home under the bridge next door. Jim was harmless, he just did not ever want to live inside, and he had friends and relatives who would regularly drop off clothing and food for him.

Mike discovered that none of the homeless people who were hanging out around the farm knew anything about what we were doing. They didn't realize that we were employing and supporting low-income people from their neighborhood. As Mike talked with people he met up with at the site, he shared stories and educated them about who we are and what we are trying to accomplish. We had no idea Mike was carrying out his own neighborhood public relations campaign. Remarkably, within a short period of time after we hired him, the theft and vandalism slowed down and eventually stopped.

Mike has told me that he "can't stand doing nothing, it just makes the night so much better when I have something to do." So on his own initiative he started weeding and cleaning up and watering, doing farm jobs at night while keeping an eye on the place. When Lissa learned of his midnight farm contributions she bought him a headlamp so he could have his hands free to work.

———

During my twenties and thirties I took my winters off from farming to travel and photograph farms and farmers around the world. When I would first arrive in a foreign place everything looked new and fresh and exotic. But if I stayed in that place for more than a few days that exotic world became familiar, like wallpaper.

My experience working at Sole Food has been similar, but with a different energy and a longer adjustment period. The Downtown Eastside, with its concentration of raw suffering playing out in public view every day, was like nothing I had experienced anywhere else. As time has progressed, though, even this neighborhood no longer appears so different and disturbing. It's not that I feel I belong here, exactly, and I haven't stopped noticing the needs of the people in the neighborhood and wanting to respond to them. It's just that over time a kind of acceptance seeps in.

But in 2012, when I first met Sandra, who goes by "Seven," I experienced again what felt to me like my first day in the neighborhood. Toothless, disheveled, dressed in black with a red bandanna around her head and a lightly fevered otherworldly expression, her appearance made me struggle to see beyond my preconceptions. My own prejudices about appearances obscured my ability to perceive Seven's depth, intelligence, and a sensitivity that emerged over time. As I got to know her, I experienced an ease and comfort that I never would have established with her if I'd met her on the streets or in the parks or anywhere outside the context of these farms.

I think that on some level Seven sensed this about me. Everyone from this neighborhood must feel that they are being judged by outsiders. Seven's told me, "When you're in the DTES"—Downtown Eastside—"and you're an addict and you're mentally unhealthy, people see you at your worst, they see you chasing drugs or in an alley with a needle in your arm all dirty and skinny." And yet she has found a place to live here, and like Lyle, she has a community. "Those who live in the neighborhood," she's said, "don't hold those things against you. We're all down in the dumps; it could happen to anybody. In mainstream society

Seven

it's hidden and not so in your face. There are so many secrets. People are less judgmental in the DTES and on the streets."

I once asked Seven about the origin of her nickname. "I was living in Calgary working at a school library putting books away, and there was this child's book called *The Seven Lost Children*. It's about seven children who walked this earth and went to many places but never found a place where they belonged. So when they died the Creator put them in the heavens as stars. Most people know the constellation as Pleiades, but native culture calls them the seven lost children. Some people know me as Sandra, but I like to call myself Seven—not always belonging to this earth, but always at home in the heavens."

When we first met, Seven was an active drug user chasing and using every day and night. She was also homeless. Crack was her drug of choice; now it is crystal meth. She doesn't use as often now, and she maintains a home.

Like many of the people we work with, before she found Sole Food Seven's life was entirely focused on her habit. She panhandled to support herself, eventually going into detox not intending to quit, but simply to take a break. At that time, she was exhausted and down to less than a hundred pounds. In detox she met a natural healing practitioner and tried some homeopathic remedies that helped.

She made a decision for the first time then, she has told me, to try to keep good things in her life. And from what she says, many good things she's learned and picked up since then remain in her life. Including us. "That was the same year I started working at Sole Food. So that year I got a job and an apartment. I cut down on using, focused on harm reduction, met some good people, and I got a cat."

She tells me too that the more she can get out to the farms—the better she is. And by *better*, here, Seven is referring to her mental health: "I'm bipolar and psychotic," she says. "I hear things and I see things. It comes and goes. You believe it because you hear it with your own ears,

see it with your own eyes. Some things that I see and hear are horrible—no one should have to see that—but I also see things that are so beautiful. You learn how to handle yourself in all kinds of different realms, but sometimes I lose my cool."

She finds the work relaxing. Seven especially loves weeding, which she finds peaceful. "When I'm weeding I imagine that each weed is a negative thing or thought and I'm pulling them out and discarding them. I get right into it, and I make sure I get every weed 'cause I don't want any negative thoughts." And here, in the soil, she says, "I think about my life and what I am doing with it. The farm is where I come to make choices. It's the right environment. It's safe, a getaway from the hood, where there are drunks and addicts everywhere. It's so easy to say yes to all that, it's what you're used to, it's what you do, you're a slave to it." But Sole Food, in Seven's case, and for so many of our farmers, provides a simple break from their hardships, burdens, and temptations. Now she's more inclined to do less drugs, to spend money on taking care of herself and her cat. "I just started a meal plan," she's recently said. "I want to get into shape before I go to see my family."

In October 2001, just a month after the tragedy in New York that rocked the world, I stood at a podium in front of twenty-five hundred people in an auditorium in California designed by Frank Lloyd Wright. The space was so well conceived that I felt like I was in someone's living room having a conversation with a group of friends. This was the annual gathering of the Bioneers.

The world was still in a state of raw vulnerability from the events of September, and this audience of cultural creatives was seeking answers and inspiration, something more than President George W. Bush's advice to the nation to get down to Disney World or go shopping. They were looking for something to hold on to that would stabilize their world, a vision for moving forward in a time of fear and uncertainty. The energy in the room was electric.

Patriotism and security were major topics in the weeks and months after 9/11, with most of the public conversation focused on finding the perpetrators, protecting our borders, and passing legislation that expanded the powers of law enforcement.

A few of us had different ideas, and the events of that time forced us to consider what security and patriotism really meant. In times of crisis it is sometimes helpful for a nation to consider its roots and the basic principles it was built on. America had been a nation of farmers, and one of its founders, Thomas Jefferson, believed that the health of the democracy was inextricably tied to the health and vitality of its agriculture. In 2001 both were in trouble.

A lot of self-proclaimed patriots waved flags and made speeches, but the real patriots may have been those who were quietly working on the land, growing food for a nation. And real security might have more to do with jobs, fertile soil, and a stable food supply than tighter borders, smarter weapons, and more surveillance.

At the turn of the millennium, the demographics of the world had just tipped from rural to predominantly urban, and many of us were wondering how we were going to sustainably feed these growing urban populations.

Rural farms and industrial agriculture were not doing a very good job of keeping us or the land healthy and well fed. The problems surfaced daily in graphic detail: epidemic levels of diabetes and obesity; groundwater pollution; soil depletion; the environmental and personal health impacts of pesticides, herbicides, and chemical fertilizers; the concerns around genetically modified food; and on and on.

We were all seeking answers about how to move forward. Urban agriculture was high on a long list of solutions, but what did it really mean? My talk that day looked at these and other issues swirling around food. I hoped to expose some of the illusions and significant challenges that now surrounded urban agriculture, and in my speech I tried to present concrete questions and ideas to make urban agriculture more real and more possible. The main feature of my talk was an *urban food manifesto*—which I've since adapted and include in chapter 10—which triggered some interesting conversations and a little controversy. It even motivated a few politicians, city planners, and activists to roll up their sleeves and try some new approaches. But deep down I knew that these ideas only danced around the edges of the real problem.

We'd all gotten hung up on individual problems, and we'd lost sight of the larger overriding cultural dilemma, the one fundamental reality that is the source of all the problems in food and agriculture—that there were so few of us growing the world's basic nourishment. The responsibility of this thing we call food, this most essential human need, had been handed over to an industrial system more concerned about the returns to its shareholders than about the personal, social, and cultural health of those whose lives and well-being depended on it. Securing food had become a peripheral activity done by a mere 2 percent of the population we called farmers.

Many people under age thirty do not know how to use their hands for anything other than pushing keys on a keyboard. The revolution can be talked about online, but we cannot produce real nourishment online. I used to say that chefs had received almost mythical rock 'n' roll status, and that it was time for farmers to receive that same attention. But the real shift we need cannot take place when only a few of us are doing the work to grow the food for the rest, while everyone else is rooting us on. I love the attention, but farming is not a spectator sport.

"There is not so much a food crisis," I proclaimed that day in California, "as there is a crisis in participation." The audience erupted when I proposed that in memory of the thousands of people who lost their lives at the World Trade Center, a portion of Ground Zero should be converted to an urban farm replete with greenhouses and kitchens and an education center. This farm would provide food and jobs year-round to those in need and become a model of a local agricultural-based economy situated on the grounds of what used to be a monument to the global economy.

I proposed the idea metaphorically rather than with any expectation of serious consideration, but the response that day clearly demonstrated

Outstanding in the Field

El cortito—the short one—referred to the short-handled hoe that was the required tool for Mexican farmworkers in California's fields during the 1960s and early 1970s. That tool required workers to stoop and bend over, resulting in physical pain, psychological stress, and long-term debilitating back problems.

Truth is, *el cortito* was also a form of control. When farmworkers were bent over it was easy for the field bosses to see if anyone was standing and therefore not working. Thanks to efforts by Cesar Chavez and the United Farmworkers, the short-handled hoe was banned in 1975.

To this day most of the fruits and vegetables consumed in North America still require a significant amount of handwork to hoe and to harvest, most of which is done by people who have traveled from points south. To do this work efficiently, and still be able to stand and walk at the end of the day, requires a lot of skill and experience, not only in choosing the right tools but also in how to use your body in the fields.

During my years farming in California many of my best teachers and mentors were men and women from Mexico whose knowledge, ability, and creativity had developed over generations of agriculture. One of the things I observed when working with those farmworkers is their stance in the fields, legs shoulder-width apart, knees slightly bent; if they're working amid row crops, they have one elbow on their thigh to provide support for the whole body. This position would alternate back and forth throughout the day. Dress was long-sleeved cotton shirts, wide-brimmed hats, and sometimes a bandanna tied around the neck—all to keep the sun off.

That farm was highly diverse, some hundred different products, roughly half in tree crops and half in row crops. That diversity made it so that our field workers

how desperately people wanted to channel all their stirred-up energy into action, something more tangible than slogans and admonitions.

Remarkably, Green Guerillas in New York City decided to champion my idea, and it was officially submitted for the World Trade Center memorial design contest. The *New York Times* ran an op-ed about it, and I received hundred of emails and letters in support. In the end the memorial that was constructed was black marble and steel and concrete, protected by a huge security apparatus, a powerful reminder of the incredible loss that was sustained there, but little that reminds us of life and renewal.

never did the same thing for too long and were able to change body positions constantly, from bending over among the veggies to standing tall in fruit trees. These varied tasks allowed bodies and minds to remain flexible as they moved and shifted throughout the day.

It is not uncommon for an inexperienced person, often half my age, to volunteer to help us on the farms, and in less than an hour to complain about the pain he's experiencing or that he can't stand up straight. And while it is true that over time certain muscle groups strengthen and adjust to this work, it's more important to be aware of how you stand or bend or carry, where you place the box when harvesting, what tool you choose and how you use it.

El cortito is well suited to our setup at Sole Food, because our growing boxes are knee-high so we can tend them without having to bend over as much. Short-handled tools work better for us than the long-handled ones. But we still need to pay attention to proper ergonomics.

Farmwork is inherently repetitive. So are breathing, blinking, the beating of a heart. We don't shy away from those life functions—if we did we'd be dead. Like breathing, eating is also necessary. Someone has to farm.

Training the body to do this work is far easier than training the mind. Disciplining the mind to accept repetition, working when it's hot or dry or cold or wet, moving heavy bulk materials—all this requires an agile mind that can deal with the negative thoughts and demons that will try to take you away.

The key, I learned from my Mexican friends, is in pacing, singing, and socializing. Rarely did I ever feel that frenetic energy from them that eventually will wear you out; singing was common in the fields, and the constant banter that passed back and forth kept the mind entertained and on track. Farming as a community experience makes happy workers, and it also makes food that tastes better.

I believe that if this proposal were put forth today, it would receive an even more positive reception. Awareness around food and its place in our lives has skyrocketed in the last fifteen years. So has our awareness of the incredibly precarious nature of the system that produces our food. Phrases that some of us have been using for decades are now part of the public lexicon. Everyone seems to be talking the talk.

But talk is cheap, and while there is an overwhelming embrace of the idea of local food and agriculture, especially among foodies, there is an enormous chasm between those who eat well and locally and can afford to do so, and those who cannot. The nutritional divide has

Radish Reflections

Market growers too often view radishes as unremarkable, easy-to-grow, whatever vegetables. Their high-speed, forty-day seed-to-harvest cycle, throw-'em-in-the-ground-in-between-slower-crops flexibility make them just a sideshow. Too often they play a two-bit part that tends to feature the more flamboyant and prima donna leading melon, tomato, or pepper.

But at Sole Food, radishes are a big deal. We grow a lot of them, and our crops are proof that not all radishes are created equal. Walk the aisles at the farmers market and you'll often see similar products; walk those same aisles with a critical eye, however, and you'll see big differences in quality and variety. Taste a radish from four different farms and you'll have four very different experiences. Spiciness, texture, moisture, color, shape—all these things are influenced by different growing practices, soils, irrigation, and harvest times.

I grew up with the standard cherry-type radish put in my lunch box or waiting in a bag in the refrigerator for some salt and a quick after-school crunch. I still love those round plump red radishes, but I have lifted myself out of my monocultural, radish-is-red-and-round belief system. I now facilitate a virtual rainbow radish parade, a carnival of color and shape, texture and flavor.

And at the market, just as customers almost always ask "Is it sweet?" when referring to a tomato or pepper or an

---

never been greater; poor nutrition and hunger continue to grow at staggering rates.

So in that speech I offered my own slogan: "Make friends with a farmer—you're going to need one." I am certain that as the global industrial experiment continues to unravel, agriculture will once again return to its rightful place at the heart, the center of our society.

The Occupy movement reminded all of us of the incredible economic disparity that exists in our world, a root cause of so many societal ills. And for this I thank them. But if we are going to focus on what we support, as much as what we are against, we should occupy our land, reoccupy our soils with life and fertility, our communities with good food. We can march on Wall Street, or we can continue to work quietly and diligently in rebuilding the real economy, the one based on soil and sunlight and people working together to grow food.

ear of corn, they look down at the humble radish and ask, "Is it spicy?" If the answer is yes, they often walk away.

We grow Cherriette, Easter Egg, Red Meat or Watermelon, Pink Beauty, Crunchy Royale, purple Bravo, black Spanish Nero Tondo, long daikon, White Icicle, Green Meat, round Ping Pong, and my favorite, our signature radish—the French Breakfast.

It's amazing to me that after all the years of public and private culinary obsession, people still look at me funny when I use the words *radish* and *breakfast* in the same sentence. It is true that most of us don't think of vegetables when considering the morning meal. With our focus on sunny-side-ups or Frosted Flakes, most of us don't include radishes in our breakfast repertoire.

The D'Avignon, our choice variety of radish—strikingly beautiful, long, slender red bulbs with perfectly isolated white tips—begs for illumination when seen piled at our stand at the market.

You can pickle, roast, sauté, or shave them, but where they really shine is when they're grated into sweet butter with a few chopped radish leaves for color then spread on good bread. Or simply put them on a plate with tops intact and score the end of the roots with a knife, then dip into salt or butter. Or slice them lengthwise and put on a butter sandwich. I often wait until there is a long queue before I voice these recipes and then watch as people step out of line to grab a bunch, something moments before they would have ignored.

Those of us who are reeducating ourselves, rediscovering our place in nature, must refine our skills and diligently work to create the local and regional models. I am sure the day will come when we will be sought after, looked to for leadership and guidance, when our farms will be the living models, the repositories that keep this sacred and essential hands-on knowledge alive.

———————

I remember the call vividly. Would I come to a meeting in Vancouver? It would only require a couple of hours of your time; we just want to share an idea, draw on your experience.

I hemmed and hawed a bit, as I always do. But there was something that piqued my interest, so I agreed.

One meeting led to another and then to another until I was fully engaged, planning and designing, ordering seeds, growing seedlings, and preparing to plant that first farm.

Nearly a decade has passed since I received that first phone call that connected me with United We Can, Seann, and Sole Food. Beyond the blur of activity that comes with projects like this, I can now reflect on what we did, what worked and what did not, and what we can share that will be valuable.

I initially agreed to participate in part because of the opportunity to reach out to people who are underserved. And I agreed to participate because, at almost sixty years old, I wanted to put to the test the question of whether production farms in the city are viable.

I had been involved in urban agricultural projects, and I had visited many projects, most of which were gardenlike in scale. This time my vision demanded something bigger, not out of some illusion that bigger is better, but because this project needed to employ and feed a lot of people. A larger scale was also important if we were to develop a model that would be taken seriously. But with big ideas comes the potential for big failures, and at the time I was not sure whether we had the capacity to make this work or to sustain it once we did.

Success is difficult to measure in this work, and I know that as a farmer my expectations are that Sole Food Street Farms should try to operate like every other farm, securing income through the food that we grow and sell. And while we have not yet fully realized our financial goals, we've been enormously successful at our social goals. Many individuals whose lives had gone off the rails have been touched by this work.

But how do you value a life reanimated with new purpose, a person given hope, an abused and abandoned piece of land made abundant and nutritious?

Sole Food was never conceived as a romantic bleeding-heart attempt to save anyone. We just thought that if we could provide meaningful work, then other things might happen as well, like better housing, freedom from the public dole, a little improvement in mental and physical health. We wanted people in the neighborhood to have someplace to go each day, and something to be a part of.

Many of these things have indeed happened, and while it is difficult to know precisely what contributes to real change in individuals or in the world, we hope that our work at Sole Food has provided a little spark, some good food, a model for others to follow.

Like all work in the world, this grand experiment will eventually pass. Each lot where we now farm will eventually grow a new high-rise and find financial value far beyond what our little farms could provide. What will endure, we hope, is the internal change of folks like Kenny and Donna and Alain and Nova and Seven and so many others—their discovery that despite where they came from or what challenges they faced, they could do something remarkable, something no one thought was possible—and that they did it one bucket of compost, one handful of seeds, one parking space at a time.

CHAPTER 10

Urban Food Manifesto

I t's a cheap trick that never fails: Place a small orange orb into the mouth of every passerby at the farmers market. It's cheap because I know that each and every one of those Sungold tomatoes contains within it a message, delivered straight to the nervous system, to seek out more.

Every Sole Food staff member who has delved into crack cocaine will tell you that the drug is like an "orgasm in your brain," so addictive it can capture and hold you after one use. I'm not suggesting that Sungold tomatoes rival crack, but human brains are wired to seek pleasure, and the unusually high sugar levels of these fruits and their distinctive fruity flavor make them irresistible.

And just as I crush garlic on the pavement so that its fragrance infuses the atmosphere around the stand, subliminally pulling in customers, or stick a bunch of basil or mint under passing noses, I use this tomato like a big, sweet, seductive smile, a tempting lure into our fruit and vegetable Kasbah where other experiences—each one more pleasurable than the last—are waiting.

I'm comfortable with our shamelessness at luring folks into our market stand. It's not like we are hawking shoddy trinkets, scrawny chickens, or polyester clothing. If we entice one more unsuspecting person into our stand using a Sungold tomato, that's good for everyone. Maybe this new shopper will discover a food she's never tried before, buy some greens, find out more about who we are.

Jesus

STREET FARM

The Sungold is the product of traditional plant breeding; it is a first-generation F1 hybrid, the progeny of the marriage of two distinctly different tomato parents. You can't plant the seeds from this tomato and expect the new generation to resemble the one you started with. Some plant breeders have tried to create an open-pollinated version of Sungold, one from which you can save the seeds to replant, but nothing I have tasted comes close.

The Japanese like tomatoes that are low-acid; the Tokita Seed Company in Japan created the Sungold. They also bred the Tomatoberry, a heart-shaped, small red tomato that they describe as "So cute, it's like candies, makes everyone delight." There's also the Sungreen tomato: "Everybody think green fruit, it must be sour, but once eat it they found, 'Wooo sweet!'"

I have often railed against fruits and vegetables that have been bred for sugar content alone. Super-sweet corn is a good example of a crop that is less like eating corn, which has a long history and culture and rich flavor, and more like getting a mainline injection of sucrose. But the Sungold is more complex; it has the highest Brix (sugar levels) of any tomato, but it also has such succulent flavor and that distinctive cherry tomato explosive crunch.

I've always wanted to know who's responsible for the work of breeding this wonderful nightshade. I want to take my hat off to those people,

offer my thanks and appreciation, and acknowledge that this one small bright orange fruit is responsible for so much pleasure, so much surprise, and a certain amount of financial success for those farmers who grow it. Including Sole Food.

I have great respect for those who participate in the slow and patient art of plant variety selection and propagation. We farmers sometimes forget that our success often must be traced backward, to someone, somewhere, who probably spent years carefully growing, observing, selecting from, and breeding the seeds and plants that make us who we are—both those who grow the plants and those who eat them. This has been going on for millennia, and growing food that tastes good and is nutritionally dynamic, productive, vigorous, and reliable is the culmination of the often unacknowledged work of those who breed and improve the essential qualities of food plants.

Let me offer just one example, and one specific word of thanks. I grew a plum in California, originally bred by Luther Burbank, called the Santa Rosa. I've grown a lot of different fruits in my farming career, but like the Sungold tomato, the Santa Rosa plum stands out. The fruit has a dark purple skin, with golden-amber flesh that is, at first, intensely sweet. This sweetness is followed by a hint of tartness that bounces around in your mouth and your brain.

I wish I had known Luther Burbank (1849–1926). His contribution to many of the foods that we take for granted is substantial. The russet potato, the freestone peach, and 113 different plums, 35 fruiting cacti, 16 blackberries, 13 raspberries, 11 quinces, 10 cherries, 10 strawberries, 10 apples, 6 chestnuts, 4 grapes, 4 pears, 2 figs, 9 types of grains and grasses, 26 vegetables, and a host of ornamental plants are all part of Burbank's legacy. All of us regularly consume the products of this one man's life and work. Burbank was criticized for his poor record keeping, something I understand well. People often criticize me for the same reason. He was more interested in physical results than he was in careful record keeping, more inclined to work in his gardens than sit at a desk.

Burbank helped create a bewildering catalog of foods with his hard work. My own contributions have been much more modest. And yet there are a few things I've helped create that I'm unreservedly proud of. Not all of these have I been able to anticipate. Who knew what else, besides perhaps good food and good work, might come of our experiment at Sole Food?

For me, one of the great surprises of farming in the city is what happens when you establish some cultivated wildness. While farms in boxes on pavement are far from nature's glory, they do provide habitat for diverse soil life, birds, insects, and larger creatures as well. Sole Food has not built up the biodiversity Burbank did, but we've done our humble best, it turns out.

At our farm on Pacific and Carrall, during the winter of 2013 the section of arugula and mustard greens that was well past its prime was eight feet tall and going to seed. Unless you visited that section of boxes on a regular basis, you would not have noticed the duck that had established her nest and was incubating her eggs in the safety of the towering plants. The nesting duck became a source of great excitement for our staff and generated all kinds of questions and conversations. What is the appropriate relationship between the cultivated and the wild within our urban farms? How close can we get to the nesting duck while working on the vegetables without upsetting her? How do we manage the crops in boxes nearby?

We checked on the duck every day, and the compassion and worry on the part of our staff was touching. I've lived around ducks before; most of them, to their knowledge, had not.

Then one day she was gone, with no remnant of the eggs. We suspect that her babies hatched and she marched them off to nearby False Creek.

The duck became the first in a series of creatures who discovered the farms and the refuge that radishes or lettuce or carrots provided from the urban environment. But not all the creatures who found their way into the farm and set up shop were as considerate and non-invasive as the duck.

Canada geese are beautiful and emblematic of this grand northern country. They are also major pests, especially when they get comfortable, give up their migratory habits, and become full-time residents.

At first the geese that visited the farm were viewed as "cute" and nonthreatening. But when staff would arrive to harvest a crop for a restaurant order and discover it mowed down by geese, or when goose poop was left covering pathways and growing beds, attitudes changed. A few of our crew feel that every creature has a right to eat, but as the crop loss and hassle became acute, everyone was enlisted in chasing geese. Eventually the increase of vehicle traffic on the remainder of the lot next to the farm deterred the geese, which resolved the problem.

The killdeer that was nesting on one of our growing boxes provided endless entertainment. The well-disguised eggs looked like stones and were laid in a simple indentation in the soil. If approached, the killdeer would move away from the nest and pretend to have a broken wing to draw predators (in this case us) toward itself and away from the nest. We witnessed this and heard its unique distress call on many occasions.

Then there was our coyote. Lissa was the first one to spot it. She'd just arrived on the False Creek farm for the day's work, and there he was, she says, "lying in one of the beds of carrots asleep." She had seen paw prints around the farm and had assumed they were those of a stray dog. But this day, the identity of the print maker was unmistakable: "I pointed the coyote out to our staff, and everyone was both excited and a little scared. He would wake up and look around and go back to sleep. The staff was worried about whether he was injured or if he got his leg caught in a rat trap. So we made noise and tried to get him to move. The coyote stood up, looking not the slightest bit concerned, yawned, and laid back down again."

He eventually just disappeared.

We once had a baby gray seagull at the Astoria farm that couldn't fly. He stayed for about a month. We also provided a home to a river otter. It was late October 2014, and things had smelled foul around the False Creek farm for some time, but we didn't know why. Then we discovered the otter hanging out under the office, leaving piles of nasty scat on a pile of lumber. Next it moved under the tarp where we store soil amendments.

And so the farms have become a refuge for wild creatures, as well as a refuge for humans who need meaningful work and a break from the harsh realities of their lives in the city, beyond our fences. It's been an important lesson for our staff that, as farmers, it is also our job to provide a safe environment for a whole world of living things we can't see—the millions of tiny organisms that are the engines of healthy soil. I often wish that I could see into, and show the staff, that invisible microbiological world; to observe the biodynamic landscapes beneath our feet would be amazing.

There are animals that have made the Sole Food farms their home, and while they all have been uninvited, their presence has been at best entertaining and educational, if sometimes a little odoriferous. But there is one creature that I regard as the bane of our existence. The rat. Farming in the city, I've developed personal hatred for rats.

Like most cities, Vancouver has a rodent problem in general. The back alleys of the Downtown Eastside are swarming with rats.

It took a couple of years for the rats to discover our plots, but once they did, word spread. The farms became a beacon for every rodent that learned there were fresh tomatoes and peppers and lettuce and beans and melons and kale growing in their neighborhood, just blocks away.

Turns out all the finest garbage in the world will never hold up to the artisan-quality fruits and vegetables we grow. The Downtown Eastside rat population has now become so sophisticated about food that they

choose only the lacinato kale, leaving alone less exciting cultivars; they devour the haricots verts, leaving untouched the more common and mundane Blue Lake beans. They seek out only the largest, most beautiful, fully colored Italian Corno di Toro and thick-walled orange gourmet peppers, ignoring anything green or small or inferior.

The rats have an uncanny ability to know just when a crop is about to be harvested, moving in hordes to eat all or, in some cases, just a nibble of every beet or carrot or turnip. I am certain that the rats that have discovered our farms are eating a far healthier diet than many of the human inhabitants of this city. In 2013 we estimated our losses to rodent damage in the range of $20,000.

We have tried everything to control rodents on the farms: hundreds of snap traps baited with Kraft Extra Smooth Peanut Butter, which we purchased in an industrial-sized pack from Costco; sticky traps; cornmeal mixed with plaster of paris, water traps—*everything*!

It is difficult to openly discuss this problem without making people uncomfortable about the food on offer at Sole Food, but rodents are a hard reality for anyone who is growing food in any city. They are a fact of urban life, and while we may be able to control them, we will never eradicate them.

Farming in the city on parking lots requires incessant attention to the efficient use of space. Space in the city is valuable, it is expensive, and it is limited, especially if we're talking horizontal. Yet for the most part, there is always room to grow up, to grow vertically; indeed, sky-high could be possible if there were not power lines and airplanes.

And so, as part of our constant reenvisioning of Sole Food, we attempted our own version of vertical production. We installed our first upright farm experiment in February 2011, on the corner of Georgia and Hornby in front of the Vancouver Art Gallery, in the heart of the shopping and business district of the city. The throngs here are well dressed, sophisticated, and purposeful as they make their way to and from offices or stores or venture out from the comfort of their rooms at the nearby Georgia, Four Seasons, or Fairmont hotels.

Our garden-scale version of one of our farms was part of We Vancouver, an exhibit that was curated for the Vancouver Art Gallery featuring innovative projects that are informing the future of the city. The exhibit included work by architects, clothing designers, writers, engineers, a bicycle fabricator, musicians, and farmers—that is, us. The problem was that the exhibit opened in mid-February, and it is not so easy to grow anything outdoors and have it appear robust and healthy in Vancouver at that time of year.

Several weeks before the exhibit was to open, Seann, Rob, Kenny, Alain, and I got started. I had never tried a vertical production system, and there were no premade devices that I was aware of that would accomplish our vertical goals, and so we made our own, not really knowing whether they would work. We had decided to work with ten-foot-long, six-inch-diameter PVC pipes as growing towers. First we drilled a pattern of holes along the entire length of the pipes.. We planned to dig holes to install the pipes in an open unlandscaped plot of land in front of the museum. What we did not know was that our installation was directly above the underground vault that holds the museum's stored art collection. While digging the holes, we encountered the first layers of moisture barrier of the roof over that vault. A staff member of the museum, whom I had summoned after our breakthrough, bundled up to come outside to see what we had encountered. Controlling his panic, he stopped us just short of penetrating that roof.

After installing the pipes, a process that took several days, we filled them with soil and planted eight varieties of kale. Kale was one of the few things that would survive February's cold conditions. As the plants grew, they filled out up and down and around the poles that were placed alongside beds of pink-, red-, gold-, green-, white-, and purple-stemmed rainbow chard. The view was arresting. Shoppers and businesspeople stopped to gawk in wonder at the display. Here was our garden, flourishing in front of an art gallery in the peak of winter, growing vertically on poles we had brightly painted with Day-Glo colors extending high into the air. The effect was disorienting, and that's part of Sole Food's mission: to shake us all out of our conventional ideas of where food is grown, how it is grown, and who grows it, and to remind us that even right here, and even right now, in February in British Columbia, such a bounty is possible.

The art gallery kale was our first sojourn into vertical growing. The experiment was not enough of a trial for us to work out all the details, but it did inspire us to try growing strawberries in towers at the Astoria farm.

The following month, a team of six people planted four thousand strawberry plants in hundreds of vertical pipes we set up and attached to the ends of the wooden growing boxes using steel plumber's strapping. These pipes surrounded the entire half-acre Astoria site, and we calculated that if we had planted those four thousand vertical plants "on the

flat," in the beds, it would have required the whole lot. Our skyward planting had essentially doubled the production area of the site.

Strawberries grow well hanging from the sides of pipes. There is sun exposure and airflow around the plants, it is warmer up and off the ground, and the fruit remains clean and is easy to harvest. Our challenge, however, was hydrological: We were unable to maintain even moisture in the soil through the full length of the pipes. We watered from the top of the pipes, and we also situated aluminum turkey-basting pans under the pipes to allow the soil to wick moisture from below. We contemplated running drip irrigation lines through the centers of the pipes, but in the end the experiment stalled and failed, not so much because it did not have potential, but because our expansion to our other sites dominated our attention. We hope to try again another season.

———————

In the fall of 2014 we put up a large tent in the center of the orchard at Main and Terminal and invited all of the organizations and individuals who had supported us to an evening dinner. Bankers, developers, foundations, individual donors, chefs, city government representatives, and the mayor all gathered together with Nova, Alain, Lissa, Donna, Lyle, Seann, and me for an evening dinner prepared with our food by some of Vancouver's top culinary talents.

Here we were, once again a strange and incongruent sight: white tablecloths, good wine and beer, beautiful food and people, and Main Street traffic zipping by, the Golden Arches looming overhead, and on the other side of the fence, some of Vancouver's homeless lining up to get a sleeping mat and a little floor space at the Aboriginal Friendship Centre shelter.

I started off the evening by leading a tour of the orchard. I recalled that some of those in attendance had doubted that we could successfully grow some of the trees and fruits that were now visibly vigorous and abundant. We smelled and tasted the multiple varieties of sages, mints, thymes, oreganos, lavenders, and lemon verbenas filling the spaces between the trees. Local chefs—Wes Young from Wildebeest, Tret Jordan from Homer Street, Rachel Lovick from Aphrodite, and Dave Gunawan from Farmer's Apprentice—gave a brief introduction to the dishes they had prepared. Toward the end of the meal I presented a slide show of the other farm sites and thanked everyone present for their contributions. Seann and I spoke of the power of the Sole Food experience in changing people and places. Nova and Alain spoke eloquently about their personal experience, and the mayor led off what became a series of high-profile testimonials and subsequent offers to help.

As the evening wound down, and all the guests had departed, we cleaned up, loaded, and packed away the leftovers. When the last chair and table were gone and everyone had left I sat down on the edge of one of the growing boxes. The city was finally settling down, and traffic had diminished. Buildings were only partially illuminated, and only a few souls wandered outside the fence on the edges of the farm.

My hand slipped off the edge of the box into a patch of lemon balm. Its fragrance released, wafted up, and danced around me. And in that moment, sitting alone in a place I would normally not be at night, I felt safe and secure, and I knew the work we were doing was right.

———————

Our farmers offer us stories about how their lives have been changed by their work. In our own uniquely grounded way, we take on the social elements, the drugs and the booze, that have ruined lives. But there's no

Miranda

Let the Circle Be Unbroken

There is a universal law of reciprocity and exchange that those of us who farm see play out in intimate detail on a daily basis. We know that we cannot treat soil like an unlimited reserve; we know that every living thing we engage with requires some form of nourishment, which in turn nourishes us. Each time we harvest a box of food and send that food off from the farm, we are shipping essential and valuable soil nutrients in the form of a tomato, pepper, radish, or carrot.

Every day an armada of ships and planes and trains and trucks enters our cities carrying foods that contain the embodied soil energy of some distant piece of land. The waste from those foods, in the form of organic matter or fecal waste, is either hauled away to clog our landfills or flushed down toilets to eventually pollute our rivers and oceans.

Rarely is there an appropriate nutrient exchange between those who are eating food and the land from which it came.

When I was in China in the early 1980s, enterprising individuals would build public waterless "toilets" along busy thoroughfares with signage encouraging passersby to stop and make a deposit. That waste was then sold as a resource to contractors, who would compost it and pass it on to farmers. Every rural village at that time had a large community open-pit toilet where everyone squatted together to contribute to the common good of the land. Coming from the West, it took some getting used to for me to join in that public effort.

Most of the "developed" world lives with a broad case of fecal phobia. We shit in precious fresh water, then use more fresh water to flush away those

fighting nature, and each year in October and November farm production slows to the point that we have to lay people off for the winter. It is one of the weaknesses of our program, a harsh irony that during the most difficult season, when the rain and cold and clouds descend, and the days become short and dark, we have no choice but to separate our farmers from the stability and community of work. The loss of income is not the only hardship they experience; they also lose the support system that the farms provide, at the worst time of year. Those of us who are in management often don't know where our famers go during those months, where they spend their time, or how they fare day to day. Sole Food is all the more reason to look ahead to spring. "It is the unfortunate valuable nutrients. We spend billions of dollars to "clean" it up so that it then can be disposed of.

An essential and critical life-giving circle has been broken. Most farms that feed people are located far from where those people live, and the industrial system that grows and distributes those foods is based on synthesized and chemical nutrients and a one-way delivery system where the only thing returning to the land is money to buy more fertilizer. Organic farms are an improvement, but unless they are dealing directly with those they are feeding, their nutrient cycle can be one-directional as well.

We need to envision our farms as self-sustaining living organisms that produce the majority of soil nutrients from within, but until we can realize that lofty goal we need to find a way to create a more appropriate nutrient exchange.

Urban farms are better positioned to take advantage of the waste being generated in their neighborhoods, but urban farms are equally challenged with the lack of space, infrastructure, and the permits to properly transport, process, and use that waste.

Every city should be separating and converting all of its organic waste and making usable compost for both urban and peri-urban farms. Most large-scale urban composting programs are not stringent enough in separating waste or in the conversion of that waste for their compost to be safe for use in food production.

All of us, whether we realize it or not, are inextricably tied to the soil that nourishes us. There are international campaigns to protect wildlife, oceans, rivers, air, and forests, but how often do you hear concerns expressed about that thin layer of material that covers our earth—the earth's placenta, which feeds and nourishes every living thing? Let the circle be unbroken.

reality in our current system at Sole Food that we cannot support all of our staff with work through the winter," Lissa has told me. She says, and I agree, that "it seems cruel that in the darkest, coldest, dampest months of the year our sometimes-homeless, sometimes-depressed staff are left to their own devices."

From our point of view as managers and directors, the winter layoffs have made us question whether our model is working. If we cannot provide wintertime employment for those who need it, are we really fulfilling our goals? It is true that the northwestern growing season provides a limiting overlay that is out of our control, but if we are to balance our social goals with our farming goals, must we be so fixated and reliant on the sales of our products? We know we can raise funds to fill the gaps—shouldn't we rely on that support to continue paying staff through the winter? These are my questions, but Lissa has posed others.

Lissa asks: On an urban farm, where costs of living are higher, how can we afford to pay a living wage and still sell a generally undervalued product? How can we balance the wage/income equation when our workplace is based more on compassion and personal growth than efficiency?

Though we have begun to explore and establish related enterprises that can provide winter employment, we're not there yet. We dabbled unsuccessfully in a permanent retail store, and we considered a compost-making operation, but although these ventures would provide winter employment, we do not have the capacity to support them. We dream of a Sole Food building with a commercial kitchen where we could can and dehydrate and freeze our raw farm products to diversify our market offerings. The space could also serve as a gathering place for our staff and farm community, and we could create a simple café that would focus on foods we grow at our farm sites.

When I ponder these larger wintertime questions, I'm still not sure I have the answers. And when spring does arrive, it's not all warmth and new growth. I am reminded of the wagons that used to tour midwestern rural areas in the late 1800s picking up individuals who had gone insane after a long isolated winter. Putting our staff back together in the spring is not unlike this; we pick up the pieces and attempt to re-create a team. And while some staff are happy to have a break after the intensity of a long farm season, the winter break is too long and allows many folks to fall back into old patterns, addictions, and habits.

But we refuse to give up on ourselves or our farmers. And like me, Lissa seems inclined to keep working, to look ahead: "These questions will probably stick with us for a while," she's said. "So for now, it's enough to bunch the carrots and put in winter starts and cover seeds with row cover and talk about next season."

———

Throughout this book, I've told a story of building a system of urban farms, employing people from the neighborhood, and envisioning a future where people and the soil are the centerpiece of our considerations of how food comes to us. I've explained that in every step along the way in creating Sole Food, and in building up farms over the last four decades, I've been guided by a vision of justice and a longing for good food. The planning I did with Seann many years ago, like the planning I do today, took shape around compelling questions, many of which seem rhetorical, some of which have found answers. New questions come up every day. Here are the questions that I am grappling with now.

- How do we grow food and feed our communities without depleting and destroying our soils, polluting our water, and making ourselves sick?

- How do we create a food system that is more appropriately and geographically integrated with a world population that is now predominantly urban?

- How do we address our society's incredible disparity in food security? Could working with the land and growing food provide disenfranchised people, especially youth, a greater sense of purpose, hope, and community responsibility?

- How do we work within a private property structure where farmable land is owned and controlled by people who, for the most part, do not engage in farming?

- How do we grow food within the ecological limitations of where we grow it?

- How do we create a food system that can absorb and survive the unpredictability of a climate that is clearly changing? And what role can food growers play in helping mitigate climate change?

- How do we balance our need to preserve and protect farmland with the reality that it is far easier to save land than it is to save farmers? What good is a protected farm without someone to work and steward it?

- How do we inspire the next generations to want to work on the land?

- How do we provide access to land and ensure that farming will be a viable way of life for those new farmers?

- In the future, whose hands will do the field work of growing and harvesting our food?

- How do farmers survive financially while helping to ensure that everyone has access to fresh food?

- How do we work within the laws and the attitudes of a nation that guards its borders to keep out the very people whose hands help grow our food?

- In the United States and Canada, where we devote a smaller percentage of our income to food than any other nations in the world, how do we educate the public that cheap food is an illusion, and that we pay for our food in many less obvious ways than the dollars paid at the checkout counter?

- What will it mean to be a farmer in coming years?

———————

I have been developing the following Urban Food Manifesto over the last ten years. Some of the ideas may sound radical; others will likely seem terribly obvious. Some are practical, some more ideological, but either way they are focused on the municipal and on individual ways to address what I consider to be some of the most prominent challenges in how we feed ourselves.

Every municipality should establish publicly supported agricultural training centers in central and accessible locations. I'm not talking about think tanks or demonstration gardens. I'm talking about working urban farms that model not only the social, cultural, and ecological benefits of farming in the city, but the economic benefits as well. We can talk about all of the wonderful reasons to farm in urban areas, but until we can demonstrate that it's possible to make a decent living doing it, it's going to be a tough sell.

Regular folks are now so removed from the work of farming that they need to literally see what's possible. They need access to those who have maintained this knowledge and those who are serious and active practitioners. Every city should have teams of trained farm advisers in numbers proportionate to the population devoted to urban food production. Those agents should operate out of their local urban agriculture centers to run training workshops and classes; they should also venture out into the community to provide on-site technical support in production, in marketing, and in food processing and preparation.

The nutrient cycle that once tied farms with those they supplied has been interrupted. We need a full-cycle food system that allows for the return of organic waste via central regional composting facilities that can support the nutrient needs of both urban farms and farms on the fringes of our urban centers. Every community could be composting all its cardboard, paper, old clothing, shoes, restaurant and grocery store waste, and on and on. We need to reduce what comes into our communities from elsewhere, but we also need to reduce what leaves those communities, especially if it has nutritional or soil conditioning values for our land.

My fields at Foxglove Farm have as many rocks as grains of soil. Removing those rocks represents a huge amount of work for me, but each one of those rocks also represents an enormous amount of embodied energy, if I could just release it. Every community should own a portable rock grinder that could be taken to farms and used to grind rocks in and around fields that contain essential minerals now being mined elsewhere at great ecological cost. There are huge holes in the world, entire mountains removed, to supply minerals such as gypsum and lime and rock phosphate to our farms. We cannot talk about a sustainable agriculture unless we address where the minerals—especially phosphorous—are going to come from.

We've all heard about peak oil; we need to prepare for peak water and peak phosphorous. We can grow food without oil, but we cannot grow it without phosphorous and water. Phosphorous is a mined mineral, which now has limited reserves, most of which are located in China, Morocco, and the Western Sahara. Some scientists believe that at the rate we now use it, remaining reserves will be depleted within fifty to one hundred years.

Let's get over our phobia around human waste, stop spending billions of dollars to flush it away and pollute our rivers and oceans, and start recycling it onto our farms and gardens. Urine is the best local source of phosphorous, and we need to figure out creative ways to recycle it.

Every community should support the construction and funding of a permanent covered year-round farmers market space in a dominant central location. Providing this type of physical space is just as important to our civic health, if not more, as the public swimming pool, the sports fields, schools, churches, and libraries.

Every new permit for a housing development should be contingent on inclusion of an approved food-production component on a scale relative to the number of people who will live in the development. And every new office or retail building should be engineered for a full-scale rooftop food production component, including greenhouses warmed by the spent heat vented from the building.

Every neighborhood school and church should be required to restructure existing institutional-kitchen facilities to accommodate cooperative canning, freezing, and dehydrating services for their neighborhoods during non-peak hours.

Every real estate transaction should include a small urban farmland preservation tax from which lands could be purchased specifically for the production of food, and those lands could have protective easements that require agricultural use in perpetuity.

A great deal of privately owned arable land currently lies fallow. This land could be made available to new farmers under long-term leases. We need to recognize that there is not necessarily any relationship between landownership and land stewardship. The only requirement for landownership in our society is access to capital. That's not enough. I believe that ownership of land should come with a set of responsibilities.

Building inspections are common practice prior to many real estate transactions; we should require land inspections, including ecological assessments and baseline documentation, on every piece of land over five acres. Every land purchaser should be required to attend a stewardship and restoration training course based on the particularities of that piece of land. This will help move land away from its status as commodity and bring some sense of stewardship into ownership.

When I was in school my favorite classes were wood shop, metal shop, mechanics, and home economics, which included cooking and sewing. Those subjects were well respected. I looked forward to shop class far more than math or science or English. It was a time when I could make something real and tangible. (Every wood shop teacher I've known was missing a finger or two, and I am sure that was a requirement for those positions. I made the connection very quickly between those missing fingers and the machines we worked with.) Life skills classes are coming back into schools, but we need to give farming and cooking and mechanics and plumbing and carpentry the same status and attention as math or English or the sciences.

It sounds radical, but in the future full-time professional farmers may no longer have the luxury of raising fruits and vegetables. This should become the responsibility of individuals and families to grow for themselves in their front and backyards, on their balconies and rooftops, and in community garden plots. We could probably survive without another carrot or tomato, but we cannot live without grains and beans and protein sources.

Every municipality should initiate a phase-out of all home lawns—effective immediately—but they must also provide neighborhood training programs and technical support for home- and building owners to replace those lawns with food production.

It may be that along with growing food, the real work of farmers in the future should be seen as the sequestration of water and carbon. Anyone who has land, or is managing land, has a huge opportunity and a responsibility to address two of our greatest global challenges—water and climate. Slowing and spreading surface water and allowing it to percolate and not run off, along with learning to use land and improve soils to store and hold carbon, are urgent and essential roles that farmers need to play now and into the future.

————

Seann has now moved on to start his own farm. Donna is now working as a janitor. Alain is off social assistance for the first time in fifteen years and works as the lead supervisor for our False Creek farm. Rob has been traveling the world with a new friend. Kenny is back in rehab, steadfast in his desire to never give up on giving up heroin. Seven has finally gotten her new set of teeth. Nova is taking care of her son, Jack, and in a stable relationship. Jordan is working full-time as a poker dealer at the Edgewater casino that borders one of our farms. Cam is learning new skills every day and becoming a good farmer, and Lyle is off morphine for the first time since the jump that broke his feet.

Midway through our most recent growing season, the staff decided they wanted to take all of Sole Food's seconds and give them away to other members of their neighborhood.

So every Sunday throughout the season, they collect the food left over from the prior week's markets and walk the five blocks to the corner of Hastings and Carrall, where the Downtown Eastside neighborhood street market is held. Typically, they share a tent and table with the hot-dog vendor and hand out beets, lettuce, carrots, peppers, tomatoes, berries, and whatever else is available at the time.

One Sunday, Lyle and I met up at the False Creek farm to bring the produce to the market. Unfortunately, on that day, none of my keys to the Sole Food kingdom provide access to the walk-in cooler, so we have to give up on distributing food and walk empty-handed to the market together. Along the way kids are playing soccer in the park, a group of men hang out nearby drinking beer and smoking, and at the corner of Pender and Carrall a man wrapped in a blanket sleeps on the sidewalk next to several shopping carts full of belongings.

When we reach the street market a skinny woman, her head bent unnaturally to the side and arms shaking uncontrollably, stops to receive a hug from Lyle. Just past the hug, like some gift from the heavens, Lyle finds a brand-new Shop-Vac still in its original unopened box just sitting on the street.

Lyle and I stand by, watching the circuslike parade of people and stuff, and eventually we decide to move on. Lyle turns as if to confide in me. Then he asks me for some money for his daily dose of heroin and reassures me that he only uses enough to keep

Lyle

away the jitters, not to get high. Each and every day he does this hustle, begging or borrowing or participating in whatever is required to take care of the monkey on his back.

As we part company Lyle stops, leans toward me, and tells me that he and Kenny ran into each other at a bus stop recently. Both had just scored heroin, and both were on their way to use it. Now, his face looks sad and so tired as he says, "There was a moment when I looked at Kenny and I knew without any doubt that we both had the same thought—'What the fuck are we doing?'"

With Kenny, and thanks to Sole Food, Lyle had a "tender moment," he says. Then he turns and walks away, Shop-Vac on his shoulder and $30 in his pocket.

Acknowledgments

This book required a whole community to make happen. I want to thank the following people: Seann Dory for collaborating with me on creating Sole Food, Lissa Goldstein for her exceptional work on the farms and with our staff, and Kelsey Brick and Tabitha Mcloughlin for their contributions to Sole Food's marketing and administration. Steve Beck, Ann Wisehart, and Jeanne-Marie Herman for reading and providing feedback on early drafts of the book. Jerilyn Hesse, Steve Beck, Sydney Ocean, and Sean and Dorie Hutchinson for housing, feeding, and taking care of me during my escapes from home to write. Frank Giustra, whose council and major funding was essential in Sole Food's early development and expansion. Terry Hui and Olivia Lam for funding and helping design and manufacture our growing boxes, and for supporting our vision early on. Vancity for its unwavering financial support, its sound guidance, and for all that it does in the Vancouver community and beyond. Ken and Judy Spencer; Nancy, Alan, and Rebecca Berman; Sue and Wieland Wettstein; Rory Holland; Rob and Ruth Peters; Andres and Michael Dean; Timothy Kendrick; Russell Precious; Mickey McLeod; British Columbia Automobile Association; the Social Venture partners; John Swift and the Swift Foundation; Ken Vidalin; the Central City Foundation; the Vancouver Foundation; Salt Spring Coffee; and the Real Estate Foundation of British Columbia for their generous support in funding Sole Food. Andy Broderick for his steadfast support and valuable advice. Terry Hui, Matt Meehan, Elizabeth Allegretto, and Peter Udzenija of Concord Pacific for generously providing the valuable space that houses our largest farm. Kira Gerwing for her work on our behalf within the city of Vancouver's bureaucracy. Mayor Gregor Robertson and City Manager Sadhu Johnston for their leadership and vision in Vancouver and for believing in Sole Food and allowing it to develop and thrive. Scott Korb for his invaluable input on the structure of the book. Margo Baldwin, Ben Watson, and Fern Marshall Bradley at Chelsea Green for their patience with the challenges and delays of my writing process and for their work in publishing the book. My wife, Jeanne-Marie, and my sons, Benjamin and Aaron, for their immeasurable love and patience during yet another long journey of making a book; my mother for her love and energy; my father, who passed on during the completion of the manuscript; and the entire crew at Sole Food past and present.

Farming Principles and Practices

When I give presentations about Sole Food Street Farms, the urban farmers and would-be urban farmers in the audience often have questions about the details of how we set up our growing system, particularly what kind of soil and containers we use and how we propagate our plants. So here I describe some of the experiments we've tried and the system we've refined for our farms. For good measure, I also include a list of important farming principles and some of my thoughts on deciding what to grow and how to market what you grow.

Foundations

Our modern existence is fraught with endless paradoxes. Every time I pour a bath, turn on the lights, or eat a meal I am participating in a vast ripple of repercussions. I try to make choices that have the lowest impact, but I also accept the duality of my existence as I navigate life's daily contradictions while holding on to a larger social and environmental vision.

Soil is the foundation of what we do as farmers. We are all inextricably tied to it, and we all depend on it. On my rural open-field farm I work with the native soil that is already there and try to make it better. Farming on pavement or on soils that are too contaminated is complicated.

At Sole Food we've had to accept the financial and ecological costs of accommodating a farming system that attempts to produce commercial quantities of food on pavement in boxes. We know that we cannot safely grow food in contaminated urban soils or physically grow on impervious surfaces like pavement, and from the start we had to acknowledge the imperfections of the system.

Our original plan at Sole Food was to make soil by composting Vancouver's waste organic materials. But an urban soil-making operation requires an enormous amount of physical and fossil energy. It involves moving vast amounts of raw materials, screening and separating compost, blending the raw materials, and then distributing the final product. We shelved our plan after realizing how hard it would be to meet permit requirements, find a site that would allow waste recycling, and support the venture financially and logistically. Thus we have had to rely on purchased and donated soils that are made locally.

Soil in boxes is separated from that broader environment. Biological relationships are limited in a box, and while we try to infuse that biology into the boxes, our efforts will always be imperfect.

We discovered the hard way that many standard practices used in open-field agriculture, such as cover cropping, were far more difficult, and in some cases impossible, to implement in urban container-based systems. There were basic physical challenges in incorporating crop residues and green manure cover crops into containers.

We had trouble implementing good crop rotations because we didn't have the space and could not rely on an extensive diversity within an urban production model. We also discovered that shallow soil depth, and the dramatic soil temperature fluctuations that were occurring in our raised containers, were problematic for some crops. Containers also limited our ability to use machinery for tasks such as tillage and mowing, and the impacts of the hand tillage we did seemed to have a more dramatic negative effect on soil structure.

On the positive side, containers have virtually no soil compaction, excellent drainage (assuming the original design and installation were correct), earlier production, and more rapid fruition (because more heat reaches the raised root zone). Boxes also have the physical and psychological advantages for farmers who can work crops that are higher off the ground and require less bending over.

We continue to experiment with different ideas and approaches to address these fundamental challenges, always believing that we will eventually evolve a more efficient, more ecologically sensible, and more economical approach to maintaining good soil fertility in large-scale box agriculture. But while it is far better than food traveling long distances from industrial farms, it will always be an imperfect system established to feed and employ folks who live in an equally imperfect urban construct.

The fundamental philosophy of feeding the soil and not the plant is also more difficult to adhere to when growing plants in containers. The urban environment presents an unlimited supply of raw materials for making compost, but the regulations, health and odor concerns, difficulty in finding a suitable location, and challenges of handling large volumes of organic materials make composting a nearly impossible endeavor on a significant scale in the city. Enclosed self-contained composting systems are a great option, but the capital cost of manufactured units is high—prohibitively so for a start-up farm project.

The early years growing in containers at Sole Food, with new rich loose soil, provided wonderful results. Plants and leaves and roots and fruit grew with abandon. Those raised boxes received heat from the sides and were well drained. The tilth in the boxes remained loose and open, all of which contributed to ideal conditions for plant health.

But after a couple of seasons I could see things changing. The organic matter content of those boxed soils was disappearing, leaving behind a soil structure that was more like powder. Despite the worm castings and compost and mineral fertilizers we added, it was difficult to maintain the soil biology and fertility within the contained and constricted environment of a box.

Our current system combines the use of concentrated mineral and powdered organic fertilizer mixes and compost tea. We use compost tea because soil fertility is about much more than chemistry; soil biology is the engine that drives the system. Using compost tea extracts provides a far more efficient alternative to applying gross volumes of compost. Extracting, transporting, and applying microorganisms requires far fewer resources and energizes the soil. Ratio-feed injectors are wonderful tools for this purpose; driven purely on water pressure, they inject minute amounts of compost tea into irrigation lines with every watering.

We also rely on minerals and other powdered amendments in a dry concentrated fertilizer mix that we apply before each planting. This mix may include glacial rock dust, greensand, rock phosphate, alfalfa meal, bonemeal, blood meal, mined potassium sulfate, lignite, kelp meal, and gypsum.

Every agricultural system has ongoing ecological compromises; our urban agricultural system is full of them. All so-called natural fertilizers are either mined or harvested from some distant environment at varying levels of ecological, financial, and, in some cases, social expense. While every farm, rural or urban, faces the reality of having to amend with mineral fertilizers, our ultimate goal should always be to view the farm as a self-sustaining, self-feeding living organism. Creating a fertility cycle from within to support the health of the farm organism is far more achievable on a rural farm that can support a mix of animal and plant life and the kinds of longer-term rotations that are necessary. Farming in boxes on parking lots limits that possibility.

Containers

High land values, contaminated soils, and pavement are core challenges for farmers in most urban areas. And while there are situations that allow for growing in native soils, they are the exception. With these considerations, we designed our container growing system.

In-ground growing is far less expensive, requiring no costly containers and imported soil, but the responsibility associated with ensuring that native soils and the food that comes from them are not contaminated is huge. Clean urban soils are rare, and it is difficult to be sure that contamination pockets (even in relatively clean areas) do not exist. Even in those situations where native soil is clean, the investment in improving the fertility of that soil may not make financial sense unless the lease is long-term.

In pursuing the perfect container for growing food we went through an amazing number of ideas, possibilities, and design and construction incarnations. As is often the case, cost became a primary driver during the early phase of our project, so we initially looked at existing containers that could be repurposed, such as plastic bulb boxes from Holland that farmers use for storing and transporting produce. While there are millions of them out in the world, we determined that they were too small. We considered collapsible, waxed cardboard watermelon and winter squash bins, but they are too deep and not durable over time. We had prototype containers built using heavy fiber material from the collapsed roof of a Vancouver sports facility. We tried fruit bins, large bags, pots, and pipes. We eventually realized that having boxes fabricated to our specifications was the best route.

After experimenting with dimensional lumber, which was too expensive on any significant scale, eventually decomposed, and was difficult to move, we settled on pallet collars for our initial site expansion on our largest production site.

Unfortunately untreated plywood pallet collars do not hold up to the elements. Not only did ours started delaminating within the first year, but rats also took up residence in the pallets underneath, feeding on our vegetables. The financial and production hit we took on this large-scale failure was huge and resulted in a decision to invest in more expensive, well-designed plastic containers as replacements and for new site expansions.

While the boxes on our largest vegetable site were unraveling, we began building our one-acre urban orchard using plastic boxes. These containers were twice the depth of our vegetable containers to accommodate the deeper rooting requirement of the trees. We designed these with a lip around all four edges to support hoops for covering beds with plastic or row cover, and forklift tabs underneath to allow for complete movability.

The boxes for the trees were manufactured locally and are virtually indestructible. I tried to destroy one with a Bobcat tractor by running over it with the crawler's tracks, crushing it with the bucket, and applying every form of abuse I could come up with. The box survived, did not crack or break, and was put back into service after regaining (with some help) its original shape.

We've also replaced the three thousand original pallet collar vegetable boxes on our largest site with plastic.

The initial "premium" plastic vegetable growing box we designed for the False Creek site had every bell and whistle—it was double-walled to provide insulation to prevent excessive soil heat and had a reservoir for recycling spent irrigation water—a feature we had to abandon because of the cost. We included tabs to allow the boxes to be stackable when full and nestable when empty, forklift slots, and a lip around the top with holes for supporting hoops. We also included scored holes in these boxes to allow for interconnected drainage so that spent irrigation water could be collected and recycled.

Propagation

There is no place where time and space play out as profoundly as in the propagation stage of our work. The smaller the farm, the more critical it is to have a working propagation program that produces high-quality transplants. This allows the farmer to get a head start growing a new batch of plants while the crop that it will replace is still completing its productive cycle in the field.

This is an especially important piece in any urban agriculture application, because the growing areas tend to be small. Not all crops are transplanted—some will be seeded directly—but well-grown transplants provide a significant boost in time and quality for any small farm.

The most common systems for propagating plants for transplant use soil blocks and wood and plastic flats, which come in an array of sizes. We use plastic "speedling"-type flats that are sized based on the needed transplanting time, germination rates, and growth rates of each crop. We use a high-density flat of four hundred cells, for example, with a crop like parsley that has a long germination time and requires a high number of transplants because of close field spacing. The smaller the cell size, the faster that cell will fill with roots and be ready to move into the fields. For most of our crops we use two-hundred-cell flats. This size ensures fast plant removal and transplanting in the field.

Other crops like tomatoes or peppers will start their lives in small-celled flats and then be moved to larger cells in order to produce transplants that are further developed. This also allows cold-sensitive crops to take longer advantage of the controlled protected atmosphere of a propagation facility.

A simple potting mix recipe for filling flats is one part sand, two parts compost, one part worm castings, one part peat moss or coco peat, and a sprinkling of seaweed concentrate or very well-aged manure, or blood or feather meal. Mix very thoroughly. Because of the relatively large volume of transplants we produce, we have found that it can be far less expensive and time consuming to purchase an organic mix than to produce our own. The costs and availability will vary widely depending on where you are, how many transplants you produce, and other factors.

The moisture level of the mix prior to putting it into the flats is critical. The mix should feel moist, but it should not form clumps or feel soggy or wet.

We make the mix in a large cart. To fill the flats, we place them in the cart and rub the mix across the flat without worry about losing any excess. We make sure every cell is packed full and then scrape the surface of the flat level.

We fill all the flats we need for that day's seeding and place an empty flat on top, lining it up with the cells before pressing down in order to leave small indentations in the center of each cell that accommodate the seed. Prior to seeding, we place a label in the corner of each flat with the date and cultivar. Seed is dropped directly from packets, tapping the packet so that only a single seed rolls out into each cell. When the flat is seeded we line it up on the top of the growing bench and cover the newly seeded flats with a thin layer of mix.

The first irrigation must be done carefully. If the mix was too dry seeds can float to the surface, and if watering is not done with a gentle spray seeds and soil mix can be washed from the flat.

Attention to well-timed and careful watering is critical to success. Timing of watering is dictated by the volume of the cells, ambient air temperatures and humidity, and exposure to sunlight. Flats should never dry out, but they should also never be soggy, so find a balance.

Transplants tend to use up available nutrients in small cells or pots quickly. Occasionally we use a seaweed drench to water transplants, especially long-term transplants such as peppers or tomatoes that grow in flats for a long time.

Transplants are ready for planting outdoors when you can gently pull them by the stem from the flat with all their roots and rootball intact.

Not all crops are field-planted from transplants. All of our root crops and cutting greens are direct-seeded. We use pinpoint seeders to direct-seed. These seeders plant at the high densities we require to achieve substantial production in the boxes, and they allow us to sow seed close to the edges of the boxes to maximize the use of limited space.

Both direct-seeding and transplanting schedules are developed in advance to eliminate guesswork in the midst of the frenzy of the work and to achieve well-timed, consistent harvests evenly spaced throughout the season.

General Farming Principles

- Details! On a small farm it's not the big ideas that lead to success, but all the little things done really well.
- Make it a priority to grow those crops you or your family like to eat.
- Maintain a consistent supply of a small body of signature products.
- Be clear about why you want to do this work.
- Design, plant, and harvest your farm on paper before a single seed is ordered or land secured.
- Likewise, know your market before you plant the first seed.
- The smaller the farm the higher the value of every leaf, every fruit, and every root.
- Crop diversity can be a form of financial security, as it allows you to ride through changing climates, conditions, and market demands.
- The more diversity you grow, the more skill you require.
- Become intimate with the land you will farm. Watch how the light moves throughout the year; study the air, the water. Discover the unique characteristics of every field, the subtle changes in soil, climate, and exposure. Know your land as intimately as you know your girlfriend, your boyfriend, lover, wife, husband.

- Find out the history of the land you are on, who came before, what they were doing on that land, and what the outcomes were.
- If you want your kids to farm, don't raise them on one.
- Limit your debt or, if possible, stay debt-free.
- Start with simple crops and narrow crop diversity, then expand and diversify as you improve your skills.

Feeding Your Neighbor, Your Farmer, and Yourself

The question of what to grow is one that requires some understanding of the community you are serving, the level of skills you have as a farmer, existing markets, local climate, scale, availability of skilled labor, and access to and cost of water.

I always advise to begin by producing those foods you like to eat; you'll naturally do a better job growing them. If you are just beginning, you may want to focus on salad greens and root crops and expand your product mix as you build your skills and confidence.

Chances are that your product mix will be dictated by economics and by space. Limited urban space requires that we choose products that will have the highest financial return from the smallest area. Some of this is informed by regional differences, but the real edge comes when you are the earliest or the latest to come to market with certain crops. Being the earliest at the market forges loyalty; being the last one to have something provides lasting loyalty. Both early and late often bring higher prices.

It is also essential to choose a small body of what I call signature crops that you become known for and that you evolve and perfect the production of. These become the crops that you make a commitment to supplying with consistent quality for the full range of the growing season. It is all too common for a less experienced grower to show up and sell a crop for a week or two, never to be seen again. Your customers want to be able to depend on you. Their loyalty will come when you can show up week after week with a dependable base body of products.

It is invaluable to informally survey local farmers markets, stores, restaurants, and CSAs and find out what is already available in your area. While it is true that you will likely find many similar products among growers, it is also true that the quality and varieties of those products will vary a great deal. Take that information, blend it with your comfort and skills, and then consider the production space you have available to create a list of six to eight items that you feel you can grow consistently over the whole season. Do not be intimidated by producing something that is already available if you think you can do it differently, more consistently, or earlier and later.

Always remember that unless you have zero financial needs, you will have to look carefully at the crop mix you select relative to its cost of production and its value in the marketplace. It is likely that the cost of production in the city (due to infrastructure and labor costs) will be higher, so consider that as you choose your product mix. What products can you grow in relatively small spaces that will provide the highest yield and the highest financial return per square foot?

I also make a practice of visualizing my display tables at the farmers market or my availability lists to restaurants when I am ordering seeds or producing a crop list. What are the relationships of the crops I am choosing to one another visually, in their culinary use, and in terms of rotation, soil fertility, or space demands? Imagine your display and how it will look while making this list and while ordering seeds. How can you stand out at the farmers market or in the eyes of a chef? What can you do that will be unique?

Direct-Marketing Principles

- Pile it high and watch it fly.
- Sample, sample, and sample some more (have a full-time sampler and give away lots of food).
- Come to market with crops early and late.
- Alternate products with contrasting colors on the display.
- Every product should delineate itself visually on the display, and every individual bunch or head or bag should be easy to identify from the whole.
- Bring more food than you think you can sell (food usually does not sell well if left in the field or in the cooler).
- Be dependable—always have a body of signature products available throughout the season that customers reliably can get from you week after week.
- Provide a no-questions-asked guarantee for any product you sell.
- Honor loyalty with discounts to regular customers.
- Always have a regular, recognizable, likable face at the market.
- Farming is sexy; use that to your advantage.
- Don't choose direct marketing if you are not outgoing.
- Be outrageous, theatrical, and have fun while marketing—it will sell more food.

- Never sit down; be upright and face-to-face.
- Never stop moving, restacking, and managing the display.
- Maintain a visual sense of abundance throughout the day even when products have diminished.
- The direct marketing of food requires energy and a constant buzz around your stall.
- Never be in a hurry to disperse the crowds. A line is good, the more the merrier, as long as people are entertained and not waiting too long.
- Appeal to all the senses by creating a space where there are visually dynamic displays, delicious samples, and alluring fragrances.
- No customer should leave your stand having not tried something new. If it is difficult to sample on the spot, give them some and ask for feedback. They will return for more.
- Reward high-volume purchases with discounted pricing.
- Draw attention to less common products by describing in a few words how to prepare them.
- Design your market display in advance on paper.
- Allow time for a last-minute market staff meeting to go over discounts, and which are the top six products that you are there to sell. Discuss strategies for the day.
- Allot more space to those items that you need to sell more of.
- Each market will be different. Create strategies to accommodate changing products, changing pricing, weather, and season.

About the Author

Zenobia Barlow

Michael Ableman, the cofounder and director of Sole Food Street Farms, is one of the early visionaries of the urban agriculture movement. He has created high-profile urban farms in Watts, California; Goleta, California; and Vancouver, British Columbia. Ableman has also worked on and advised dozens of similar projects throughout North America and the Caribbean, and he is the founder of the nonprofit Center for Urban Agriculture. He is the subject of the award-winning PBS film *Beyond Organic*, narrated by Meryl Streep. His previous books include *From the Good Earth*, *On Good Land*, and *Fields of Plenty*. Ableman lives and farms at the 120-acre Foxglove Farm on Salt Spring Island in British Columbia.

green
press
INITIATIVE

Chelsea Green Publishing is committed to preserving ancient forests and natural resources. We elected to print this title on paper containing at least 10% postconsumer recycled paper, processed chlorine-free. As a result, for this printing, we have saved:

17 Trees (40' tall and 6-8" diameter)
7,787 Gallons of Wastewater
7 million BTUs Total Energy
521 Pounds of Solid Waste
1,435 Pounds of Greenhouse Gases

Chelsea Green Publishing made this paper choice because we are a member of the Green Press Initiative, a nonprofit program dedicated to supporting authors, publishers, and suppliers in their efforts to reduce their use of fiber obtained from endangered forests. For more information, visit www.greenpressinitiative.org.

Environmental impact estimates were made using the Environmental Defense Paper Calculator. For more information visit: www.papercalculator.org.

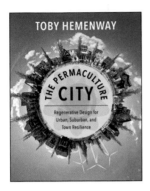

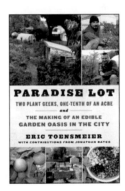

All proceeds from the Vancouver edition
of this book go to Sole Food Street Farms.

COURTESY OF THE

SpencerCreo
Foundation